Critical Thinking

Critical Thinking
Learning from Mistakes and How to Prevent Them

Gerald J. Watson Jr.

with illustrations by
Jesse J. Derouin

CRC Press
Taylor & Francis Group
Boca Raton London New York

CRC Press is an imprint of the
Taylor & Francis Group, an **Informa** business

First edition published 2021
by CRC Press
6000 Broken Sound Parkway NW, Suite 300, Boca Raton, FL 33487-2742

and by CRC Press
2 Park Square, Milton Park, Abingdon, Oxon, OX14 4RN

Library of Congress Cataloging-in-Publication Data
Names: Watson, Gerald J., Jr., author. |
Title: Critical thinking : learning from mistakes and how to prevent them /
Gerald J. Watson Jr.
Description: First edition. | Boca Raton : CRC Press, 2020. | Includes
bibliographical references and index.
Identifiers: LCCN 2020005866 (print) | LCCN 2020005867 (ebook) |
ISBN 9780367354602 (hardback) | ISBN 9780429342998 (ebook)
Subjects: LCSH: Decision making. | Critical thinking. | Errors. |
Task analysis. | Environmental psychology.
Classification: LCC HD30.23 .W375 2020 (print) | LCC HD30.23 (ebook) |
DDC 658.4/03—dc23
LC record available at https://lccn.loc.gov/2020005866
LC ebook record available at https://lccn.loc.gov/2020005867

ISBN: 978-0-367-35460-2 (hbk)
ISBN: 978-0-429-34299-8 (ebk)

Typeset in Times
by codeMantra

This book is dedicated to my wife who helped me get out of the Georgia Institute of Technology, not once but twice. She then provided moral and financial support when I returned at age 59 to work toward a PhD in Industrial and Systems Engineering at the North Carolina Agricultural and Technical University in Greensboro, which I earned at the age of 64.

It is also dedicated to many students over the years who have helped me become a better person and instructor.

Contents

Preface

What makes this book unique is that it is a compilation of errors that have been personally committed, or witnessed, by me and my close associates. Hands-on, real causes for those errors were then assigned and categorized to enable one to examine a situation that requires a decision to be made and then asked, "Is the decision maker really making the correct decision? Did the decision maker apply critical thinking to determine the worst scenario that can occur if the incorrect decision is made?"

My definition of critical thinking is "the disciplined mental activity of evaluating arguments or propositions that can guide the development of beliefs and taking action". I then delineate what I consider to be important characteristics of critical thinking: (1) something vital for success in the 21st century; (2) its concept must be separated from the concepts of good and creative thinking; (3) its associated expected behaviors and subtasks as well as functional definitions must be developed; (4) its task analysis as well as intermediate goals and means of identifying ways of achieving objectives need to be completed; and, finally, (5) its best practices for delivering critical instruction need to be developed and implemented [1].

I challenge each decision maker to use critical thinking before making a decision. Then for each decision, list an alternative and use critical thinking to determine, what is the worst thing that can occur if this decision is implemented? Is the worst scenario a situation that is acceptable? If not, then the decision maker must reevaluate the alternatives and ask, could a better decision have been made with better information? If so, how long would it have taken to get that information and at what cost? All costs must be included, such as life, dollars, and time.

MY MOTIVATION

In addition to the overall goal of using critical thinking to learn from mistakes, I had several other goals for writing this book. The overall purpose is to reduce the frequency rate and severity that result from the commission of an error. I hope to achieve this by categorizing the main causes of these errors, which have been committed by me and others. Each reason will be explained in detail to enable everyone to select a motivation that best fits the most recently committed error. I have committed at least one error in each category.

Another goal is expose the reader to the concepts that will result in fewer mistakes made and, for those that do occur, a reduction in their severity. These concepts include mostly those that I practiced as an industrial engineer during my long career; I was familiar with some of these concept names, but not all. I had taken several courses in Scientific Management and was aware of its overall goal but had little experience in its application other than conducting time studies. I had taken some practical application in usability courses primarily during layout modifications, and more application in human factors engineering primarily when applying work simplification prior to conducting time studies, but I was not aware of the correct nomenclature. I was aware of the benefits of anthropometry as I was able to sit and work at a desk but was clueless as to the proper field of study, much less its correct spelling.

Sufficient information concerning each topic is provided subsequently to enable the reader to understand how the knowledge concerning that particular concept could have prevented that mistake or, at the very minimum, mitigated the consequences. More details concerning each topic are provided at the end of the book under Interesting Topics to enable the reader to increase their knowledge concerning that particular subject.

One of the earlier courses I took toward my PhD was in Usability. The term *usability* itself is focused on the ease with which users use or interact with a product or a system. It focuses on the user, his satisfaction, efficiency, and effectiveness. Properly applied, usability engineering can help reduce the number of mistakes that are committed often by eliminating the opportunities for these errors or mistakes. When I turn on my computer, I wonder why the on/off button is not green or why it is not larger to facilitate its location since I am an engineer trying to make things easier. While working in MS Word, Excel, or PowerPoint where I spend the majority of my time, I am often curious as to why the cursor is not in some color other than white. Would a different color not be easier to see? Hello, interface designers!

Another aspect of human factors is that not only colors but also the contrast between different colors play significant roles that almost everyone sees but may not understand the underlying reason for. Warning signs that are seen every day are painted in solid yellow with the symbols in black. This combination is much easier for people to see further away, using a smaller font, in the driver's peripheral vision, in less-than-perfect weather condition, even with an eye disease, or the loss of vision due to aging. The best combination is black and white since it provides the greatest contrast in brightness followed by black and yellow [2].

Another application of color familiar to motorists everywhere is the traffic light. It was invented in Cleveland, OH, after the inventor witnessed an accident involving a horse-drawn carriage and car, and he was given a patent for this invention in 1923. The red light was designed to improve automobile safety [3]. Another benefit of its familiar design with red on top, yellow in the middle, and green on the bottom is it is one of the means of adapting to the inability to differentiate among colors. Another adaption technique requiring the assistance of others is the organization and labeling of furniture and clothing.

Color vision deficiency (CVD), the inability to differentiate between certain shades of color, is normally the result of a recessive gene inherited from the mother. The condition is more prevalent in males than females with approximately eight percent of males who suffer from CVD compared to one-half of one percent of females. The condition remains stable throughout one's life and does not cause additional vision problems. At this time, there is no cure. Diagnosis requires exhaustive eye examinations.

There are various forms of CVD, but the most common is red green. Those with these types possess a more difficult time differentiating between red and green depending on the darkness or lightness of the colors. Cones in the eye known as photoreceptors enable colors to be seen. These are found in the central part of the eyes, the macula, which recognize colors through their different wavelengths. Although CVD is a genetic disorder, it can also result from other factors such as diabetes, glaucoma, chronic alcoholism, multiple sclerosis, side effects from some medications, chemical exposure to fertilizer and styrene, and aging [4].

Critical thinking is essential to obtaining maximum benefits when employing usability engineering. As one who has traveled and rented automobiles frequently, I would usually attempt to fill gas on the wrong side of the car. I am sure that an extremely large number of complaints resulted in the addition of an indicator on the dashboard showing on which side of the car the gas tank is located. Also, looking for the ignition switch in the dark resulted in a light that illuminates the ignition in the dark facilitating key insertion. Also, now most Automatic Teller Machines (ATMs) will eject the card after the PIN is entered but before the transaction is completed to increase the likelihood that the ATM card is not left in the machine. In fact, I left my card and the machine emitted a beeping sound to ensure that I remove it. If the reader has not yet determined it at this point, I am one of those people who upon seeing a wet paint sign, the first thing I do is touch the surface because otherwise how does one know if the surface is actually wet or not.

Simply stated, usability focuses on the ease with which a product can be used [5]. It sounds simple, but often designers are more concerned with the looks of a product than how easily the product can be used. It is often applied to websites, but it is applicable to any product. Usability testing can be achieved with as few as three testers. Due to the upward trend in mistakes due to human error as a result of the increased technical sophistication of components, more and more emphasis must be placed on usability engineering.

The motivation that companies perform usability testing is to develop and produce more usable products. Usability testing is a simple process consisting of selecting a group of people, providing them an assignment, monitoring their progress or the lack of, and taking notes and timing the length to completion or surrender. The assignment can be on a website or the assembly of an item following written instructions. If using a website, eye tracking monitoring systems can be used to follow the path of the eyes and mouse. Then after the task is completed or the group gives up, a discussion is needed to follow up on improvements so that the task becomes easier to accomplish. Once the changes are made, another iteration of testing needs to occur with a different job assigned. This process of testing should occur until the testers agree that the product is more user-friendly.

The usability class visited a large company in the area to observe their usability lab. Students observed company employees in the lab making simple changes to their record to ensure that these minor changes could be made error free. Examples of these changes were address changes. An employee has numerous minor changes that need to be made in his record to keep the record updated as moving from Address A to Address B. The company believed that it was so much easier for the employee to change certain records than to take a form to HR and trust that someone in HR changes the database correctly within a reasonable time. This was the ultimate in usability. Providing employees with this ability would increase data accuracy and morale, and provide employees with some degree of control.

The class was assigned a project that consisted of working in groups of four to five students. The assignment consisted of designing a website, assigning a simple task to be performed on the website, and tracking the time in seconds and the number of clicks the observer made attempting to carry out the assignment. Class members were made acutely aware that they should not give up their day jobs at this point to become web designers. Then the focus of the assignment became to improve the design in order to reduce the time and the number of clicks by at least 25 percent. After several iterations, all groups attained their goals. Some groups required more attempts than others, but all crossed the finish line.

The reason that each team was not successful was that initially a few, if any, team members employed critical thinking. Each attacked the issue with the attitude that websites are a dime a dozen, a piece of cake—but when the task was taken as seriously as it should have been, it was realized that it was not going to be an easy class task. But upon analyzing the results, it became obvious that critical thinking should have occurred at the beginning of the project to ensure that the assignment was completed successfully. Although not a mistake, a valuable lesson was learned by all class members.

Human factors is the study of how humans interact with a system and its various elements to maximize the benefits. It is not about humans but about understanding their limitations. It incorporates those limitations into the design of workplaces and equipment. Human factors includes more than ergonomics; it includes cognitive functions as well. As one human factors professor explained, ergonomics covers all those areas from the neck down, whereas human factors covers everything from the top of the head down. Those in the fieldwork in all industries implement their knowledge to improve the ability of humans to work in their environment to reduce the likelihood of making an error or meeting with an accident.

Practitioners of human factors use various techniques to address all areas in the field. Their goal is to reduce the likelihood of an error and to improve the efficiency, creativity, job satisfaction, and productivity of systems involving humans and the systems with which they interact [6]. Human factors can be applied in any location where humans work. Various topics included are situation awareness, which is simply the understanding of what is going on around the operator. Another topic is task analysis, which has often evolved into hierarchical task analysis and is a sequential list of the tasks and technologies involved. As an industrial engineer, I referred to task analysis as elemental descriptions. These were written when taking a time and motion study on an operation. This included every component that an operator needed to perform and complete a job as well as to meet the quantity and quality requirements of the job. This is an important documentation for several reasons: one can only change production standards when a material or method has been changed, and unless there is a documentation on the current method and material, one cannot demonstrate that a change has been made, and the task analysis or elemental descriptions are used to determine the optimum tool and method to be used to select the scientific method to replace the old rule of thumb [7].

Elemental descriptions or task analysis includes the sequential order of all tasks as well as personal protective equipment (PPE) required for the job. Accuracy of information is crucial for accident investigation, methods improvement, and process improvements and redesign. These could be used as workplace instructions as well as training instruments for new employees or for annual safety training. They are crucial in assisting employees to achieve or to increase production goals. Cognitive task analysis is similar to task analysis, except that cognitive task analysis requires cognitive skills and demands be noted to enable performance to be measured against requirements.

During an introductory course, I was introduced to the NIOSH lifting equation which establishes an upper limit on the amount one person can lift. I was aware that such a limit existed but did not know what the amount was. The knowledge and implementation of this limit resulted in reducing fatigue and potential injuries, and improving productivity. This equation is used by safety professionals to assess the material handling risk that employees are tasked with in both lifting and lowering their daily duties to minimize the risks of musculoskeletal disorders and determine safe handling risks and guidelines. Readers are encouraged to pursue this subject further by visiting this and similar sites [8] including https://www.osha.gov/SLTC/etools/poultry/additional_material/niosh.html.

I have always been curious as to why doors often have the same widths and heights though there are many companies producing windows and doors—was there a conspiracy? Having worked for a manufacturer of office furniture, I was also curious as to the heights of desks regardless of manufacturer. The course I took also mentioned that the adjustability needed for humans to function comfortably at desks was accommodated by making the chairs vertically adjustable within a certain range.

The introductory course delved into anthropometry, the study of the measurement of the human body, which helped in the design of work stations that reduced employee fatigue, improved productivity, as well as provided many other benefits. During a course I taught in statistics, the subject of anthropometry was introduced to enable students to understand that the existing dimensions of widths and heights of doors were to accommodate 95 percent of the population, which is two standard distributions away from the mean.

Two known contributors to human errors, fatigue and lack of training, were not the causes of any of the mistakes observed directly or indirectly by me. Each manufacturing facility was operated by an experienced trained workforce. The workforce in each facility received training on a variety of subjects at least monthly, and supervision ensured that all employees took all the time allowed for breaks and lunches. During management and team meetings, management and supervisors emphasized the fact that employee safety was more important than productivity.

A primary reason for the trained workforce was that these organizations were established embracing the principles of Scientific Management. The pioneers Frederick Taylor and Frank and Lillian Gilbreth made overlapping contributions. Time studies and optimum work methods were the primary focus of Taylor. He was more interested in increasing output per person by focusing on the selection of the right person for the job and training, whereas the Gilbreths were more interested in the relationship between motion, movement, and productivity to eliminate unnecessary activities in order to determine the best work methods to increase productivity.

Taylor proposed that each job be investigated to determine the best way of performing that job as opposed to the previous rule of thumb, to hire the right employee for the job, monitor the performance of the employee and initiate training as needed, and to separate the work so that management can manage and the employee can perform the necessary tasks.

The Gilbreths were also interested in increasing productivity, but their focus was in a different direction. They would film an operation to enable them to study the activity in detail. Since they were devoted to improving employee welfare and motivation, they would decompose the operation into discrete elements so that they could possibly reduce the motions needed for an operation. The result of such an elemental breakdown was that it enabled a timekeeper to determine a time for that task and its elements to be taken for comparison purposes to track changes in productivity against changes made on the production floor.

The result was that much of their research was incorporated in quality control and assurance programs that were initiated during the 1920s. Over time, the study of ergonomics resulted from their cumulative work.

It must be stated that I am not referring to actions that involve life and death decisions. These types of decisions must be made at the time based on the circumstances, one's training, etc. No one is equipped to second guess that decision since only the decision maker and the participants are involved. A tour of Vietnam qualifies me to make this statement.

Author

Gerald J. Watson Jr. is a Vietnam veteran who graduated from the Georgia Institute of Technology with a BS and an MS in Industrial Management. He spent 30 years working with several companies in different industries as an industrial engineer, plant engineer, and safety team manager before returning to earn a PhD in Industrial Engineering from NC A&T. He has taught at the community college and university levels for over 20 years, where his main focus in teaching is to make students think and effectively communicate orally and in writing.

Introduction

BACKGROUND

Human errors exist because to err is human. Some are costly and some inconsequential. The definition of human error varies somewhat depending on the source, but all contain the statement that an error was committed involving a human who failed to achieve a desired outcome. The error may occur in either the execution or the planning stages. Plans can be improperly made but implemented correctly; it could be properly made but poorly implemented; and in the worst case, it could be both poorly made and poorly implemented [9]. Although the prevention of human error has been the subject of numerous articles and books, it persists today and the rate of commission is even increasing. In fact, one study suggests that the contribution from human errors has increased from a low of approximately 20 percent during the decade of the 1960s to more than 80 percent in the decade of the 1990s. This is more than a 400 percent increase. This increase is attributable to an upward trend in the sophistication of components and to mistakes that can and often occur due to creations caused by the designs of designers, situations exacerbated by managers, and decision makers [10].

The elimination of human error is impossible but what is the worst that can happen if an extra minute, hour, or even day is taken to rethink an idea or concept thoroughly to ensure that the correct decision is being made. An old friend who is not only a high school but also a college classmate has made numerous successful career changes over his career. Before each move he asked himself the question, "What is the worst that can happen if I accept this new position?" He has never failed in his career choices. Everyone should use their experiences as examples. Most decisions last forever – remember measure twice, cut once.

REASONS FOR MISTAKES

This book contains five major reasons for making errors. However, the underlying cause of each reason is the failure of the decision maker to use critical thinking to develop a list of questions and then honestly answer those questions before implementing the decision.

As pointed out in Chapter 6, over 80 percent of errors result from the failure of the decision maker to take some type of action. The cost of taking that action must then be evaluated against the cost resulting from the commission of the mistake. As I told my students many times, "It is not what you pay, but what you get for your money."

1 In a Hurry to Get the Task Completed

During the summer after barely surviving junior year at college, I was working at my father's gas station primarily pumping gas and changing oil. I had recently taken courses in productivity, bottleneck theory, and value stream mapping, and was thinking of how I could apply that knowledge to the current tasks. My father was in the next stall replacing a drain plug manually with a torque wrench, ensuring that the plug was tightened to the correct torque as per specifications. Aha! I thought, what a waste of time! I could use a pneumatic wrench to reduce the replacement time significantly and productivity would soar. My father looked at what I was about to do and suggested against it. I was not deterred—after all, my father was not a rising senior at an engineering college. Well, a few seconds later, I realized that I had stripped the drain plug and the use of the pneumatic wrench was probably not such a good idea. I looked sheepishly at my father who said that he would need to go and buy an oversized drain plug that I would have to pay for. Finding an oversized drain plug in pre-Internet days required many phone calls and involved a lot of travel time.

After finding and installing the plug, the next requirement was to guarantee that the oversized plug did not leak. I then explained to the car's owner what I had done and the motivation behind it. In an attempt to compensate the owner, I paid for the oil and filter change and filled the fuel tank with gas. As the car was being driven away, I realized that the time and money I thought I was saving was more than offset by looking for and obtaining the oversized drain plug. This does not even include the cost of the plug and the gas. What is worse and cannot be measured was the humiliation of committing such a mistake. After all, I had survived many hours of math, chemistry, and physics. What was missing?

Despite all the courses and associated laws in chemistry, math, and physics given at my college, my father asked whether students study Murphy's Law. I responded with questions concerning the possible originator of that law: Newton, Edison, Tesla, Galileo, or Einstein? My father responded that before implementing a decision, the decision maker must think of what can go wrong, because if it can, it will, and at the worst possible and least expected moment. This is known as Murphy's Law. According to an anonymous 50–50 rule, the 50–50–90 rule: anytime you have a 50–50 chance of getting something right, there's a 90 percent probability you'll get it wrong.

If I had thought through the process ahead of time and realized that a stripped drain plug could result in expensive, timely, and humiliating consequences that would cause more than the few seconds that it would save, then I would never have used the pneumatic wrench. Thus, I committed the mistake because I was in a hurry to get the task accomplished as I wanted to impress my father with all the knowledge I had gained concerning productivity to help justify his college expenses. I did not think critically about the negative consequences of the action that could occur and only thought about the positive results. This is the step that results in a majority of the mistakes that I had personally committed and witnessed.

During my professional career years later, I had the privilege of working as an outside employee at a plant, which hired an industrial engineering consulting company to install an incentive system for its 400 plus employees. An individual incentive system was established for each department over a period of several weeks. Anyone who did not meet the production standard within a specified period of time was subject to termination. My role as an outside contractor was to confirm the fairness of the incentive rates.

As an industrial engineer with many years of experience in the establishment of incentive rates in both union and non-union plants, I realized after several weeks of studying and analyzing the rates, that they were fair, and reported that to the management.

At a minimum, based on my experience, the mistakes made in this implementation were as follows:

1. Plant production personnel were not informed prior to the consulting group entering the plant to establish incentive rates. The incentive rates were to be established by members of the consulting group performing time and motion studies using the existing methodology.
2. No production personnel were given an explanation of the time study, its purpose, its use, and why it was being conducted. No operator was asked permission to be studied. No completed study was discussed with the operator or his supervisor for their input.
3. Neither the plant industrial engineer nor any supervisor were involved in the methodology or the establishment of the rates to enable them to answer the questions from employees during the establishment of or after the rates were established.
4. Training did not occur for those employees on or below the threshold of being terminated or those desiring to increase productivity.

These mistakes prompted the employees to discuss the feasibility of unionization and invite union representatives to hold meetings to do so. It also resulted in costly management meetings with all supervisors, with the need for an out-of-town labor attorney to have a proper conversation with employees, and consequently delayed shipments and cost overruns. These management meetings were mandatory and resulted in a lack of supervision on the production floor.

These mistakes were preventable and were the result of several factors, with the main one being the desire of the management to implement the incentive system as soon as possible in order to reap the economic benefits. The management failed to think critically through the situation to determine the worst possible outcomes. In the end, this failure cost the employees, the vendors, the community, and the company.

As an industrial engineer, I have installed several incentive systems, and, in my opinion, an incentive system is beneficial only if it is fair to the employees first and to the company second. However, before the company can begin installing a system, the employees must first become educated as to when and why this is implemented and then become consistently updated during the entire process.

I was once involved in implementing an incentive system in a sewing department of a furniture plant that had 26 employees, which required six months to implement: four to acquire the data and two to validate the data. The industrial engineering staff consisted of two full-time industrial engineers and a part-time technician. The engineers would check and ensure that each operator was trained on identical products on different days at different times to guarantee consistency of results. Time studies consisted of no fewer than 20 observations, with means and standard deviations computed to confirm all times were within two standard deviations (95 percent of the data). If not, then the study was not used in the final calculations. Studies were also conducted on two types of material to ensure that there was no statistical difference in sewing times between the two. In addition, sufficient studies were undertaken during the sewing of straight seams versus curved ones to ensure that there was no statistical difference between the two. Both studies were confirmed with a test of hypothesis at the 0.05 level of confidence using the t-test. Only studies that met those criteria were included. In addition, a trainer was hired to make sure that everyone could meet the new production standards. In the first two weeks of the new incentive system, many employees received the higher of the two wages. The trainer then worked with those employees who had not earned a higher rate and those who desired to increase their earnings.

The incentive system I installed required six months to replace the existing one in this furniture plant. The time granted for implementation was probably excessive, but the system that was being replaced was poorly designed and maintained. The plant manufactured low-end dinette furniture in various sizes and styles that were upholstered in fabric or vinyl. The company manufactured swivel chairs as well as chairs and barstools with four legs. The size of seats varied, but for some unknown reason when incentives were provided, three sizes were established: small, medium, and large. Initially, there must have been documentation as to which styles had a small seat, those with a medium, those with a large, or the dimensions that differentiated each, but over time these dimensions or the documentation were lost. So one day, a seat could be small and the next day the same seat could be medium or large depending on the mood of the supervisor. The different sizes had different pay rates under the old incentive system.

One mistake I observed was on a line that assembled vertical blinds for sliding glass doors or windows. The assembly operation consisted of six different independent operations, each consisting of one operator. When the operator completed the assigned task, a pre-glued barcoded label was attached, enabling him to earn an appropriate number of standard hours for that operation. Should that operator not have available work at any time during his shift, he noted the time started and stopped and the total time he was unable to work due to a lack of product availability. The downtime sheet of paper was submitted with his earned hours sheet daily to his supervisor. The employee's pay for that day was calculated by multiplying the earned hours, adding his downtime hours, multiplying by his hourly rate, and then summing the two.

The mistake was due to the disproportionate amount of time as a percentage of the total time the operators spent in completing the downtime report. I conducted a sampling of the work and concluded that each operator spent 40 percent of the time manually computing information on a downtime report, which equates to over three out of eight hours in a non-value-added effort. Three out of eight hours equates to over a fourth of the entire operator's workday that was not consumed productively. In terms of operators, the line had two operators who were not needed. I suggested a group incentive system, eliminating all the production staff from the beginning to shipping and relocating the line adjacent to shipping to reduce storage space. Due to the methodology of determining the labor standard, it was not possible for the initial operation to have downtime since unavoidable delay was one of the personal factors built into the standard. This was a costly mistake and occurred because of the failure to consider all aspects of the project.

An important concept in industrial engineering is value stream mapping. In a motion study of bricklayers, Fred Gilbreth stated that one of the steps in eliminating the rule of thumb and adopting and developing scientific knowledge is to eliminate all unneeded, false, slow, or useless movements [7].

Gilbreth simply explained this concept to his students: handle a product no more than it is necessary. The optimum number of times is one as the product is assembled and packaged for shipment. To prevent unforced idle time and/or bottlenecks, the assembly line must have sufficient availability of products to each employee on the assembly line to enable a continuous work flow. Each employee on the line must be trained on each task on the line to enable him to assist as needed for a continuous flow of product. I worked for a furniture company in which the engineers joked that the company wore out the furniture before it was shipped to the customer due to the excessive number of times it was handled in the plant.

One aspect of the job I enjoyed was working as a part-time salesman. This provided an opportunity to meet new people and observe different operations. Often, I was able to assist a customer in reducing their material handling costs by working with local engineers or by purchasing to reduce lead time or order quantity. I called on a huge company that produced automobile tires in an adjacent state, which was currently supplied by a competitor. The company consumed more than two truckloads of tubes per week delivered on racks with casters. Obtaining this customer would add an additional plant for the manufacturer my company serviced.

I met the purchasing agent (PA) and after reading the specifications, I quoted him a price per thousand (M), free on board (FOB) customer, and shook his hand in agreement while looking him in the eye. The meeting occurred on a Monday, and he placed an order due the following Monday. Since there were several other calls that I needed to make to other customers, I did not return to the office until that Thursday to follow up with a formal written quote.

As I was calculating the price, I realized that an error had been made in favor of the customer. I quoted a price of X, but according to the formal quote, it should have been X plus 10 percent. I cursed myself for making such a careless mistake. Then, I thought about what should be done.

Luckily, my boss, the president, was in the office. So I went to see him to discuss the situation. After all, he was the president and was paid the big bucks. When I told him about the error and the extent of it, he just looked at me and told me that this was funny. He then stated that he hired me because he thought I was smart since

I had gone to an engineering college but realized that he had made a stupid mistake in hiring me. He even reminded me that I had earned a master's degree from the engineering college. He then told me to leave his office so he could call his college buddies to tell them that his "smart" engineering college graduate made stupid mistakes like everyone else.

Needless to say, I was shocked at his reaction since at the very least I expected an unpleasant discussion, if not termination. I again reminded the president that at that price the company would lose money but mentioned that a formal written quote had not been sent to the customer. He said that did not matter and asked if I had looked the customer in the eye and shook his hand. When I replied that I had, the president responded that honesty, integrity, and respect once lost can never be regained, but money can always be recouped. When I again reminded him that the company would lose money, he responded that a deduction would be taken from my salary to cover the loss. When I told him that the salary would not cover the loss, he joked that he would simply give me a raise. When I asked how the company would compensate for discrepancies in prices in the long term, his simple reply was that when the company gets a general increase in rate of X, the company will increase their prices X plus 1 percent and eventually the difference would be eliminated. But again he emphasized that the most important thing was not money, which could easily be replaced, but honesty, integrity, and respect, which once lost were difficult, if not impossible, to regain.

No one in the company ever mentioned this simple math error. However, when I was at the corporate headquarters, I was fair game when performing calculations. Any of the executives enjoyed approaching me with a slide rule asking if I needed assistance with addition or subtraction, knowing that those two functions cannot be performed on the device.

This error was definitely avoidable. I did not take the few extra minutes to double-check the calculations nor use critical thinking to determine the results of an inaccurate quote, which I had done in such a hurry to provide it to a new customer.

Since foreign automakers were successful in selling automobiles with diesel engines in the United States, a local automaker during the latter part of the 1970s decided to create a new range of diesel engines that would not be subject to the new federal stringent emission and mileage standards that came into force in 1972 and became more and more stringent over time. The company assigned one division to achieve this while delivering the fuel economy and performance demanded by customers. Since diesel engines obtain their power using a compression ignition system, gas laws from chemistry dictate that the temperature of a gas and its volume are inversely related, provided that the volume of the gas in the combustion chamber decreases, then the temperature increases. Thus, diesel engines require higher compression than a gasoline engine. Since the bolt pattern was not changed, the head bolts stretched and the coolant could leak into the cylinders causing catastrophic, internal damage to the engine.

This mistake occurred because the company rushed to get the conversion from the gasoline engine to the diesel engine. If critical thinking had been performed, it would have resulted in proper research that would have recommended a different head bolt pattern to withstand the higher pressure needed for compression in diesel engines. Critical thinking would also have pointed out that a common problem with diesel fuel at that time was that it was often contaminated with water. The solution would have been to install a water separator, but this was omitted to save money. The absence of the water separator resulted in corrosion in the injection pumps, fuel lines, and the injectors themselves [11].

There were other issues as well. Although this diesel was produced for a limited period of time, from 1978–1985, the losses that resulted are difficult to measure. The company lost its prestige, and numerous class action lawsuits were filed: some lawsuits reimbursed owners for as much as 80 percent of the costs of a new engine, some lawsuits established Lemon Laws in all states, and the reputations of diesel engines were tarnished for buyers at least temporarily.

These mistakes resulted from the failure to understand the importance and significance of human factors in the workplace. I encourage readers to investigate this topic further, first by reading the information under Human Factors at the Appendix section of this book and through further research and coursework.

2 To Meet a Deadline

The first set of questions to ask are: Whose deadline is it? How did it originate? Is it realistic? A management tool referred to as the Program Evaluation and Review Technique (PERT) can be and is often used to schedule, organize, and ensure that tasks are accomplished. This technique actually originated in 1958 with the development of the Polaris missile. The use of this method requires the listing of each activity that must be performed in the completion of a project, and equally important the time required to perform the activity from the beginning to the end of each task. The benefits of using PERT include the facilitation of decision-making as well as a reduction in the time and costs to make a decision. Each task in the project will be dependent on some tasks which can be performed in parallel with others and some sequentially.

To determine the time taken for each task, PERT uses three definitions: optimistic time, which is the least amount of time needed to accomplish a task or an activity; most likely time, which is the best estimate of how long it will take to accomplish a task or an activity, assuming there are no problems, or the time that has the greatest probability of occurring; and pessimistic time, which is the best estimate of the longest time that it will take to accomplish a task or an activity, assuming there will be problems. A beta distribution often used with PERT by applying a formula for the computation of the standard deviation [12] may then be written weighing these times to be used as the task time. Remember that these times are all estimates. An estimate is still an estimate. And yet, there are those who will defend these times to their last cup of coffee.

On a sales call with a salesman, I learned that a large automobile tire manufacturer produced tires that were used on airplanes in a plant several states away from their closest one. This plant was serviced by a competitor whose plant was very close which made a visit to that plant even more appealing. After talking with the PA there, I discovered that the company used four paper tubes in each corner of a corrugated container to strengthen the boxes so that containers could be stacked three high in a railroad boxcar for shipment. The type of tube used was spiral—think of the tube on which paper towels are wound—because it has some crush strength, side-to-side and end-to-end strength. The other type of paper tube is similar to a paper straw—in this case, called a convolute tube, which has more end-to-end strength and little side-to-side strength. Convolute tubes are easier and less expensive to manufacture than spiral tubes because fewer processes are involved.

While speaking with the PA, I discovered that the next order was due in one week. The self-imposed arbitrary deadline which was to receive the next full order to secure this account for the company drove me to bypass the procedures. Normal procedures were to test the existing product, provide a sample of the proposed product as well as one that equaled the currently used product with a delivery price for each, and then let the customer perform a test to determine which product performed best.

I had established a self-imposed deadline and was not going to be deterred. To meet this arbitrary deadline, he should have taken the following actions but failed to do so:

1. Obtain customer specifications for the existing tube.
2. Obtain sample tubes for testing in his company's laboratory.
3. Provide a sample of both tubes for customer testing.

On returning to the office, I determined the mileage, calculated the freight upcharge, and added that to calculate a delivery price and provided it over the phone. He was pleasantly surprised when the PA told him that the quoted price was very attractive. At that point, he should have suggested that a sample of 50 tubes or so be sent for an evaluation. But he was so eager to earn an "attaboy" that he took an order for a truckload to be delivered the following week. In the sales meeting the following week, several people congratulated me for the good job.

Unfortunately, that feeling of euphoria did not last very long. The cartons in the railroad cars were often tossed around during the shipment and the spiral tubes which were able to endure the side-to-side pressure enabled the cartons to remain intact. Conversely, the convolute tubes were not able to withstand the side-to-side pressure resulting in the cartons collapsing and releasing the airplane tires within the railroad cars. This required that the railroad cars had to be unloaded manually as opposed to the use of automatic equipment such as fork trucks. Telephone calls from unhappy union stewards and other railroad and company employees at all hours resulted in disappointment.

This costly mistake was preventable had I asked simple questions: (1) Why was the customer using the product he was using? (2) How long had he been using it? and (3) Was there any test results with the current or previous product(s)? He did not follow the steps in Engineering 101. Simply stated, he did not critically think through the situation to determine what could go wrong and act to prevent it.

These mistakes resulted from the failure to understand the importance and significance of human factors in the workplace. One such human factor is pressure: whether self-imposed or from the environment. In this situation, the pressure was self-imposed.

3 Failure to Consider All Costs of the Operation

Several examples are provided that demonstrate the mistake of failure to consider all costs of an operation. A raw material may be purchased from a non-local vendor and the freight is not included in the purchase price for some reason. The raw material could be a new experimental product with a high waste factor that is not included in the purchase price. There could be unforeseen situations that occur during manufacturing or processing, which are not included in the pricing. It is for this reason I recommend that critical thinking should be employed before any process change to determine any potential problem that may occur to prevent unforeseen costs or delays to enable a solution.

The first real job I obtained after graduating from college was working for a company that produced modular kitchen cabinets and vanities. Included were oven and utility cabinets that were not produced until an order was received because of the difficulty of the production, assembly, finishing, and packaging of these items and also their size and the number of unique components.

The company produced to inventory for several reasons: first, it had operated in that mode for a long time. Although this method of manufacturing had several advantages, it also has its disadvantages. The advantages were it enabled industrial engineers to minimize setup cost per piece to almost zero by scheduling a large batch of parts to be produced (thinking was that the parts would be used eventually) and it enabled PAs to minimize raw material costs by purchasing large quantities of raw material at a time (thinking was that the material would be used eventually). Second, the company had just constructed a large expensive warehouse for finished goods that had to be used, and sale of items was the function of the marketing and the sales department. Disadvantages included the cost of the extra raw material inventory, the likelihood that the raw material inventory would be damaged or become obsolete, and the payment of insurance and taxes on the inventory itself and the building to accommodate the inventory. The most important disadvantage was the color stain on the finished cabinets; although packaged in a corrugated container in the warehouse the color would change due to weather conditions over time. "Last in, first out" was used to schedule shipments.

Oven and utility cabinets, being labor-intensive and difficult to produce due to their size and unique components, were produced only as ordered, and could be shipped with base and wall cabinets that were produced a year or more earlier. Upon installation, the installers and customers realized the stain mismatch and the company was forced to install new recently manufactured base and wall cabinets at no cost to the customer to match the newly produced oven and/or utility cabinets.

During the 1960s and 1970s, International Business Machines (IBM) began producing computer models, the 1440 and the 360, that enabled material requirements planning (MRP) to become a reality. IBM offered new software MAPICS that offered MRP, a system that enabled a manufacturing company to produce to order instead of producing to inventory as it had done previously. This was achieved by exploding (decomposing) a bill of materials into its individual components, separating dependent demand from independent demand based on coding, calculating the need for lower-level inputs (independent demand) and either alerting the PA or placing an order as needed [13]. This enabled the assembly, sanding department, and finishing room to finish a complete kitchen at one time. Inspection of a complete kitchen or bathroom occurred before packing and shipment. This eliminated complaints due to stain mismatch since all cabinets were assembled, sanded, and stained at the same time. In the machine room, mass production was undertaken on common parts and minimum production on unique parts to ensure that the entire order could be produced together.

This costly mistake because of a stain mismatch was easily avoidable and occurred as a result of lack of critical thinking. When a company maintains a large finished goods inventory, the cost of excess inventory must include the cost of maintaining the inventory, the tax on the inventory, the potential damage that can occur, the possibility of obsolescence of the inventory due to changes in the market or the material itself, and the loss of interest on invested capital. These costs must be summed and compared to the benefits of just-in-time production and inventory systems.

Another costly mistake occurred as a result of the purchase of a computer numerically controlled (CNC) fabric cutter. The machine was to be used in a plant located at a small rural town in the middle of nowhere. The closest airport was at least a two-hour drive depending on traffic. Due to the price difference of approximately 25K between the one made overseas and the one produced in the United States, it was decided to purchase the machine produced overseas. The machine was received but did not meet the US electrical requirements as specified in the purchase order but met the requirements of the country in which the fabric cutter was manufactured. Also, the technical manuals for the fabric cutter were in the language of the country in which the cutter was manufactured, which in this case was Spanish. Guess how many people in the rural small town in the middle of nowhere knew Spanish—none.

The first thing that had to be corrected was the electrical requirements. Since there was no locals who knew Spanish, the company had to find an electrician who knew Spanish. Needless to say, there were not too many local electricians, and those who did exist had to come from a distant community and charged much for their time and travel.

The machine was finally cutting fabric to everyone's satisfaction. Then there was a malfunction, and since the technical manuals were in Spanish, the maintenance people in the plant had no clue as to how to make necessary repairs. A telephone call to the manufacturer was essentially useless because of language differences. Maintenance personnel from the manufacturing company were flown over to make repairs. Of course, the repairs were not covered under warranty because of shipment—whatever excuse can be contrived—so the difference in price was becoming less and less. Of course, the maintenance personnel needed to stay in first-class accommodations, and vehicles and meals provided during their stay further reduced the savings. Eventually, the maintenance personnel left and returned home.

Due to the maintenance department's inability to make repairs as needed, the maintenance crew from the manufacturer became almost regular but extremely expensive visitors. When the machine was not cutting fabric, the plant had to resort to the previous costly practice of manually cutting fabric.

This mistake was definitely avoidable. The purchaser did not use critical thinking to evaluate all potential problems that could result. It did not take many round trips for several people from Spain and accommodations plus lost production to pay for the difference in the two machines. One cannot simply look at the purchase cost but must consider all the factors involved.

The mistake was not due to the machine being foreign-made. It was due to the failure to critically think about the situation to determine what could go wrong with the purchase, particularly when it was foreign-made. The primary things that could go and did go wrong were problems related to maintenance and repair parts since downtime on the machine seriously affected productivity and costs.

Several failures acting together resulted in this mistake. One was the failure to consider all aspects of the project. Questions that should have been asked initially were the following: (1) Why is the machine less costly than its competitors? (2) Where else is this machine being used? (3) What about training and maintenance on this equipment since it is foreign-made?

Another was the failure to ask for advice. This CNC fabric cutter was new technology for this facility in the middle of nowhere. Maintenance needed to be involved initially in the decision-making in order to establish a preventative maintenance program, to understand its operation and to have its manuals translated into English. The availability of technicians and repair parts for all equipment regardless of its durability was not considered. If it was the only machine that was to be in the United States, then the decision should have received additional scrutiny because training and acquisition of repair parts would become much more difficult.

Another possible failure was a human factor and the pressure from the management due to potential cost savings. This facility was facing extreme pressure to reduce costs due to competition and the use of this fabric cutter would save direct labor and material costs by a reduction in waste. At that time, fabric was the most expensive cost of furniture and a reduction in its costs combined with a reduction in direct labor would provide the competitive edge needed.

This mistake is a combination of several failures: not considering all aspects of the project, failure to ask others for assistance and pressures from outside. However, the failure most responsible for this costly mistake was the failure to take all aspects of the project into consideration. If this had been done, other factors would not have been significant. The motivating factor in this decision was the huge price difference between the two machines. As I had told my students many times, it is not what you pay but what you get for your money.

To emphasize the validity of that statement, I had a company car in the early 1970s with a slant 6 engine with an automatic transmission. It seemed as if the engine was on life support but someone forgot to provide the necessary power. It did not look forward to its daily activities. As a result, it rewarded me by its sluggish performance and turning into a gas station of its own accord. Acceleration from 0 to 60 required a day, so it was amazing that more rear-end collisions did not occur when entering interstates. Mechanics at the shop knew that I was frugal, a nice word for cheap, and challenged him to use unleaded supreme gas. Their challenge was that the use of gasoline over regular unleaded would result in better performance and mileage. Being skeptical and curious at the same time, I took on their challenge especially since he had a certain brand of a company credit card and did not personally pay for gas. This was the only change.

In the first month, I used a certain brand of regular unleaded gas at various prices, traveling about 4,000 miles a month. During that month, I maintained meticulous records and used a certain brand of an unleaded supreme during the second month, again traveling about 4,000 miles keeping detailed records paying various prices. At the first of each month, oil and filter were changed, tires were rotated, and air checked. The only thing that changed from one month to the next was the type of gas used. The cost of gas per mile was identical to the fourth decimal point, proving that despite the higher cost per gallon, the variability in the cost per gallon, and the gas mileage was increased enough to offset the higher cost per gallon.

This experiment confirms in practice what I knew in theory and has conveyed to students, fellow employees, and essentially anyone who will listen. It is not what you pay but what you get for your money.

On a late Friday afternoon, I had a job interview with the plant manager of a furniture plant that had been in business forever. The plant produced a traditional line of bedroom and dining room furniture. After speaking with him for a few minutes, the plant manager told me that he was going to save bazillions by reducing 25 percent of the material costs. I asked how he was going to achieve this feat. He responded that he was going to replace all the unseen components of wood in furniture from whatever was being currently used to sweet gum. Sweet gum was less than half the price per thousand board feet of the least expensive grade of lumber he was using at the time.

Several weeks before the job interview, I had paid a tree removal company an amount resembling the national debt to remove a sweet gum tree next to his house and cut it up into two-foot sections so that it could be split for firewood. I was unable to split it, despite weighing 160 pounds of pure muscle. With a log splitter powered with a zillion horsepower Cummings diesel engine, I do not think I would have experienced much success. I asked the plant manager if he had ever tried splitting sweet gum and of course his response was no. He then told me that engineers that he spoke with at various woodworking machine companies indicated that he should not have a problem. I asked if he was going to try it on a small scale just to make sure that it would work as he anticipated. His response was that he had not. What was I thinking?

In addition to the difficulty in splitting, other problems exist with sweet gum. Although there is ample supply, little is seen in sawmills or furniture plants. One problem is that it shrinks or warps in all directions because the material has interlocked grains [14].

This costly mistake was entirely preventable had the plant manager employed critical thinking to question the cause that sweet gum was half the price of the least expensive grade of lumber. This additional waste would more than compensate for the lower price. As I advised my students, it is not what you pay but what you get for your money. Most of the time, you can pay somewhat more and save money in the long run. As much as possible one must always be aware of incremental costs to enable comparison with incremental revenue.

One plant in which I had the opportunity to consult produced vertical window/door coverings used to primarily cover sliding glass doors. This company purchased plastic material in eight-foot lengths into which various items were inserted, such as fabric and metal.

Slots for hanging above the sliding glass doors were then punched into these inserts to enable drawstrings to be attached for the closure of the coverings. The median length of the material sold was 84 inches with a standard deviation of exactly 1 inch. The raw material was priced and purchased per linear foot.

The process consisted of receiving the raw material in the warehouse, retrieving it by one person, and bringing it to the first station in the assembly line, which is the cutting table, cutting to length and discarding the waste (mean of 12 inches into an adjacent trash container and paying for waste disposal), punching one end to accept the plastic insert and drawstring, moving it to the next workstation that attached the plastic inserts into the vanes, placing it onto a conveyor to enable assemblies to be added at the next workstation inside the headstock that enabled the vanes to rotate vertically 180°, returning it to the conveyor as needed to enable quality assurance (QA) to perform an inspection to guarantee that the vane quantity was correct and the mechanism inside the head assembly functioned, and finally packaging the product at the next station for shipment.

The amount of vane material purchased exceeded 100K linear feet per year. The PA advised me that he was in the process of converting from purchasing 8–12 linear feet lengths and that he was going to save the company an extraordinary amount of money. At first, I thought he was joking but when I realized the PA was serious, I tried to make him realize when several factors added together, the purchase of the additional four feet resulted in the company losing money on the product. Instead of disposing of an average of one foot per blind, now the company would throw away five feet. The costs of disposal would be multiplied by a factor of five, and the raw material which could currently be handled by one person would now have to be handled by two people due to its length, doubling the costs of receiving and bringing it to the first workstation (Figures 3.1 and 3.2).

This mistake was avoidable if the PA had taken the time to consider the effects that the change would have on the manufacturing process. He did not use critical thinking to evaluate all the possible consequences of the changes. As a professor at the college, I eloquently stated and my statement can be simply paraphrased "every change causes a bump on a balloon—you must look and make sure that the bump you get is smaller than the one that was there, to begin with." This PA failed to do that. He did not critically think before he acted and thus despite my logical argument, he made a costly mistake for the company and himself. Not only did the mistake cost the company many dollars but also the loss of his job.

While working for a large office furniture company as a manufacturing engineer, one of the many responsibilities I had was to enter the bill of material, routing into the computer system and certifying its accuracy. The frame for a new product was a purchased part that was priced free on board (FOB) vendor, which means that

TITLE	Vertical Vane Material (orthographic view)			
Figure Number	3.1 Drawn By JD		Date	7/20/2019
Vertical Vane Material				
Checked by JW	Date	8/15/2019 ECO #	978-001	

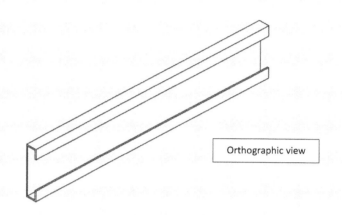

Orthographic view

FIGURE 3.1   Vertical vane material—orthogonal view.

TITLE	Vertical Vane Material (plan, top, and right-hand side views)			
Figure Number	3.2 Drawn By JD		Date	7/22/2019
Vertical Vane Material				
Checked by JW	Date	8/15/2019 ECO #	978-001	

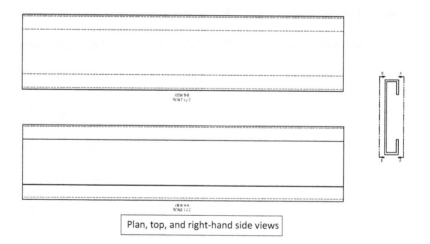

Plan, top, and right-hand side views

FIGURE 3.2   Vertical vane material—top, plan, and right-hand side views.

the customer is responsible for paying the incoming freight. The frame was powder coated with aluminum color before shipping and the quality was guaranteed to be 100 percent usable. Powder coating is a method of finishing that is applied as a free-flowing dry powder and then allowed to cure under heat to form a skin, which is a hard finish that is usually more durable than most other type of finishes.

Quality inspection of initial shipments revealed that usable products ranged from 60 to 85 percent. Inspection required 1.27 man hours per frame. I disagreed with the PA on the pricing terms and thought that the cost should be the FOB plant to enable that the purchase costs from this vendor could be easily compared with the quote from other vendors should any be located. Because the waste factor was high and varied, I used a special technique to set up the bill for the purchased frame. A bill of materials (BOM) comprises the components and quantities to manufacture or repair a product or service. The BOM consisted of the purchased frame, a waste percentage which was the mean of wastes from the most recent five shipments to include the cost of scrap, the inspection labor cost based on the waste percentage multiplied by the mean time for inspection, and the freight cost, which was the mean of the most recent cost of the last five shipments. As the vendor improved his quality, the inspection rate, freight cost, and frame cost would decrease. Using this special technique, the quality factor and freight changes were easy to track.

I discovered a local vendor who not only guaranteed 100 percent quality but also next-day delivery, FOB plant. The delivery guarantee applied regardless of the order quantity. The only problem was that the price was $14 more per frame than the existing vendor. I gathered data on incoming freight costs, quality inspection costs, costs due to disruption in production as a result of stock-outs, and long-term losses that the company would incur if the company remained with the current vendor. The only response from the PA was the vendor would improve. I responded, "What guarantee does the company have that the vendor would improve to the point that he will be as cost-effective as the one I discovered and how much money would the company lose before the existing vendor arrived at that point?"

The situation was the result of the failure of the PA to perform research to determine the vendor's capability of meeting or exceeding the required quality and quantity demands. There are several factors that contributed to this avoidable situation. First is that the PA failed to research all potential vendors and factors other than the purchase cost. At a minimum, this research should have consisted of plant tours, speaking with plant employees, discussions with existing customers, reviews with better business bureaus, chambers of commerce, etc. In my experience, more can be paid for your product if that additional cost can be justified. In this situation, the reduction in waste alone was sufficient to offset the additional cost of the frame. The additional savings that would result from the reduction in freight costs and incoming inspection costs would have added to the company's gross profit. Time could have played a role, but it did not since the product had been in development for more than a year. A factor that contributed to the mistake was the failure to ask for help from experts. The product was to be produced in what has been historically the center of the furniture industry in an existing company plant. No employee in that plant was asked for their advice or input. Worse yet is that the PA did not make a trip to the area to visit many potential vendors. A combination of numerous factors resulted in this mistake, but the overriding one was the failure to consider all costs of the operation which would have included additional waste, the additional cost of inspection, freight, and inventory.

Inspection costs and costs to return a defective product, costs of production interruptions due to stock-outs, and costs of carrying extra raw material inventory due to the variability in the quantity of acceptable order received needed to be added to the purchase cost to obtain the actual cost of the purchased frame.

Another mistake that I witnessed involved a PA concerning the item on the bottom of an office swivel chair, a caster. There are five casters per chair. At one time, there were four but these chairs easily tipped over and could no longer be produced. Changes to the materials and/or processes by which a chair was processed must be approved by all involved including marketing, purchasing, engineering, quality, and all levels of management. The process by which these changes occur is the Engineering Change Order (ECO) system; in most companies with multiple plants, this system is electronic. I received an ECO to change from the current vendor to the one in a different country. The cost difference was $0.01 per caster. The caster was priced FOB manufacturing plant, which meant that the company for which I worked was responsible for paying the freight from this foreign country to the facility I worked. This compared unfavorably with the terms of the current vendor which was FOB in this facility.

All other plants in this company signed the ECO, signifying agreement with the change. I did not sign the ECO and indicated the following reasons:

1. There was no attached documentation indicating that the caster from the new vendor was of good quality as the existing caster.
2. There was no documentation indicating that a discussion had occurred with the existing vendor to which a competitor had provided a lower quote, and the existing vendor had been given an opportunity to meet or exceed it.
3. The company I worked had been conducting business with the current vendor for many years with no quality or delivery concerns.
4. My opinion was as much manufacturing as needed to be retained in the United States as possible.

The PA was not happy with the response, but appreciated the honesty. Several weeks later, I received a call from a fellow engineer from a sister plant in another state asking if the plant at which I worked had any casters from the original vendor. I responded that the plant did but when asked why, the engineer stated that the first shipment from the alternate vendor passed inspection and performed as expected but when the cartons from the second shipment were opened, all that could be seen were plastic components. The casters had apparently broken apart in shipment. I checked with the material planner in his plant to determine how much, if any, could be shipped to help a sister plant. The material planner called the fellow engineer and told him that the plant would ship as many as possible without jeopardizing that plant's shipments. The plant at which I worked placed an order with the original vendor for as many casters as possible and deliver these at the earliest.

Saving a penny per caster seems negligible, but one must be aware that there are five casters per chair. The company produced over 6,000 chairs per day in their different chair plants, so a change was justifiable if proper conditions were met. However, this mistake resulted in the inability to ship chairs for many weeks resulting in many dollars lost and prestige which cannot be measured in dollars. I would not want to explain to a customer that his chair could not be shipped due to the desire to save a penny a caster. The only plant that did continue to produce and ship chairs without interruption was the one that did not change casters. The motivations of the plant to not approve the change were that the freight costs from the new vendor and the cost to carry the needed inventory to cover the additional travel time would exceed the penny difference in the cost of the casters.

This costly mistake could have been avoided. The PA did not include all costs of the caster and for some reason failed to use critical thinking to determine all potential possibilities that could result from this decision. Additional costs that should have been included were freight from the new vendor as well as the cost of the additional inventory that would be needed to ensure that there were no stock-outs. Once the company was unable to ship the product, lost revenue must also be added to the costs. Additionally, the company for which I worked had been purchasing from the vendor for many years without quality or delivery issues. One must treat vendors as one treats employees—with respect, honesty, and dignity.

These mistakes resulted from the failure to understand the importance and significance of human factors in the workplace. One such human factor is lack of self-assertiveness. Another contributing human factor could have been lack of team work. I encourage readers to investigate this topic further by reading the information under Human Factors in the Appendix section of this book and carrying out further research and coursework.

4 Failure to Consider All Aspects of the Project

One day a new employee of an existing contractor came to work. An existing employee of that contractor was walking him through human resources and became aggravated at the amount of time that the HR person was taking to process the new employee. After a few hours, the existing employee, who was paid hourly by the contractor, began using inappropriate language and directing it toward the HR person assisting the potential new employee.

The HR manager called me and explained the situation. I immediately called the police who worked at the facility and went to the HR office. I calmly explained to the employee who was upset that the length of time required to process a new potential employee could not be quickened by him or his language, and that his language was inappropriate for anyone, especially a female, and that his services were no longer needed as of that moment. The employee was escorted off the premises by the police to his car and followed to the gate to make sure that he departed.

His role was to take the new employee to HR to initiate processing and then report to the job site. I then called the company and explained the situation and told them that the employee had an anger management problem and that he was not welcome at the facility again. In this situation, the HR person had one task—to process the new employee, and the existing employee was to return to the job that he was hired to perform.

The first mistake committed was that the existing employee failed to realize that the HR person had a procedure that must be followed and needed that time. The second mistake committed by the existing employee was his failure to treat people. This is a lesson that I learned the hard way over many years. He failed to employ critical thinking to realize the benefits of treating fellow employees with proper dignity and respect. Failure to do so will usually result in non-optimal performance of fellow employees or worse. Treating fellow employees and those with whom one works with dignity and respect is a very important aspect of a project. As someone who had recently graduated from college, on the first day at a job in a union plant, due to an inappropriate comment made to a union steward, an unauthorized strike almost occurred until tempers moderated. The result was four hours of lost production, a very unpleasant discussion with the plant manager which almost resulted in the loss of my job after one day, which would not have been a resume enhancer, and an important lesson for me.

Another mistake I witnessed was in a project that he inherited and which involved the installation of an air handler that required a crane for installation through the roof of a four-story building. The outside dimensions of the air handler were four feet by ten feet but it was modular, meaning that it could be disassembled into two sections each four feet by five feet. The technician in charge of the project instructed the contractor to cut a section in the roof exactly four feet by five feet because that was the outside dimensions of the air handler that was to fit into the section four stories off the ground.

Since the contract had been assigned, the contractor spent an entire day attempting to fit a four feet by five feet air handler into a four feet by five feet cut-out roof section four stories off the ground to no avail. Finally, the contractor realized that it was impossible without enlarging the section, necessitating an emergency work order for the contract that had to be negotiated to increase dimensions of the cut-out so that the air handler could be inserted. Of course, the contract to cover the cut-out in the roof had to be then renegotiated since the original terms of the contract had been altered. The emergency work order had to be hand-carried through all channels of communication and authority to enable the work to continue and the monies to be appropriated and channeled from other sources. The project was delayed several days awaiting approval and funding.

This costly mistake occurred due to the lack of critical thinking and could have been avoided. It occurred due to the lack of consideration of all aspects of the project. No one questioned how an air handler with certain dimensions that is hanging from a crane and would be affected by the wind would fit into a cut-out section of a roof with the same dimensions particularly four stories off the ground.

While in high school, I worked for a Ford dealership and was given instructions by the general manager to take a fellow employee and go to another Ford dealership in another state and return with a new 1962 Thunderbird. When I returned with the car, the general manager began to have a non-positive discussion, because the car did not have all the items that were on the Sales Document posted on the window. As soon as he began, I stopped and asked him what he had instructed me to do. He responded with "To go and get this car and return it to the dealership". Then I stated that at no time did he mention that I had to check to make sure that everything was on the car that was supposed to be on the car. Needless to say, I did not go to retrieve another new car for that dealership. The only reason I did not lose my job was that my father worked at the same dealership and okayed the workmanship of all the mechanics.

This mistake resulted from the general manager assuming that I knew to check every option of the car to pick up the error and from the failure of the general manager to consider all aspects of the project.

My friend had to pay a late fee and a trucker overtime as a result of an assumption. The trucker was to make a delivery that required an overnight stay at a motel and an early departure to ensure on-time delivery. The truck driver needed to leave the motel no later than 9 am in order to meet the delivery deadline or my friend would have to pay the late fee. For some reason, my friend called the trucker at 9 am on the day the load was to be delivered to check on the status. The trucker replied that he was still at the motel. When asked why, my friend was told that the checkout time was not until 11 am. Apparently, the trucker had never stayed in a motel and did not understand that he could leave any time before the checkout time. That was an expensive lesson for my friend to learn.

Both mistakes resulted from the failure to consider all aspects of the project. One can never assume but must always be specific even to the point of redundancy. The general manager assumed that I knew to check the price list on the windshield of the car that listed all the accessories to confirm that everything that was supposed to be on the car was indeed on the car. As it turned out, there were a number of expensive items missing and the general manager had to request that the other dealer send them. Concerning my friend, it is hard to believe that not everyone has stayed in a motel but my friend learned it the hard way. In giving instructions, one cannot assume anything even knowing that checkout time means that anyone can leave before but not after that time.

As a manufacturing engineer of a large office company who processed specials, I received orders for these chairs since a unique bill of material (BOM) and routing was required to process them. A special was defined as a chair that required a non-standard component or process. It could be as simple as a customer supplying a particular fabric to one with different dimensions to accommodate a conference room. A BOM consists of all the components, both purchased and manufactured, required to manufacture one item, in this instance, a chair. A routing contains all the operations needed to manufacture, inspect, and ship the manufactured item, in this instance, a chair. These operations could be performed within the plant or by an outside contractor.

I received a large order for expensive swivel chairs in special blue leather to match the blue carpet that were to be installed in an office in a distant state. The company for which I worked ordered this leather from Sweden; the company in which the carpet was to be installed ordered carpet from Dalton, GA. The material planner, in order to save time and money, found a JPEG image of the carpet on the Internet, sent it to the vendor in Sweden along with a purchase order for enough hides to produce the number of chairs ordered. Leather was ordered in hides and was the amount of leather from a single cow. The leather arrived in the seating plant; the chairs were produced, assembled, and shipped.

This mistake occurred due to the failure of the material planner to realize and consider that just as there are color swatches used to purchase paint, the same color swatches are applied to the carpet and almost everything. In fact, the Pantone Matching System (PMS) is a color standardization system that enables everyone to identify and match colors [15].

The problem was realized because the carpet arrived first and was laid on the office floor. The chairs arrived later and although not next to the carpet, there was sufficient difference between the two blues that the chairs were not acceptable. This was a huge costly mistake particularly because this was a very large customer. In an attempt to address the problem, the material planner ordered a square yard of the carpet and had it airfreighted to Sweden. After the notification by quality assurance (QA) and the customer that the color of the hides was a match with the carpet, the material planner sent instructions to the vendor to airfreight as many hides and as soon as possible to an address in this distant state. The company I worked for sent a leather cutter, a sewer, an upholsterer, and an assembly person to this other state for the duration until all the leather on the chairs were replaced. Overtime and first-class accommodations for the crew who were sent to the other state were required. The company I worked for lost its prestige more than money for this mistake. The company also suffered from having to pay overtime to the remaining employees to maintain the schedule in the plant.

The cost of the mistake was a misguided effort to save time and money. The material planner did not employ critical thinking to prevent this costly mistake. Had the planner considered all the aspects of the project, the first thing he would have done was to acquire a square yard of the carpet to be used as well as the PMS color. Then the carpet and the PMS color should have been shipped to Sweden with instructions to match the carpet. When QA agreed that the colors matched, then the order for the hides should have been shipped. This error occurred because the material planner had not considered all aspects of the project—the need for two different blues to match and the only method of assurance was by the PMS color standard.

The first project I was assigned in my first job was to layout the finished goods warehouse. Being a recent graduate, I wanted to impress my boss, who was also an alumnus of the college I attended. The warehouse was a large (250,000 sq. feet) two-story pre-engineered type building. I gathered sales data, square feet data, cubic feet data, etc. computed the quantity of each item to maintain an inventory, determined locations in the warehouse to minimize travel distance to dock doors based on items most used, and used linear programming to determine optimum location size and placement for each item to minimize travel distance (time) to the dock doors. I patted myself on the back for a job I thought had been done well.

My boss looked at the layout and asked one question—"Where are the posts"? I responded that they were only 4 inches square and the warehouse was almost 6 acres and that the posts surely could not get in the way. The next Saturday morning, I had four temporary laborers meet at 7 am so that I could get an early start. As the first aisle was marked with blue chalk, the posts could not have been more in the middle even if they had planned for them to be. My boss, who naturally had to witness the situation, only said: "He was glad that the posts are not going to be in the way". Disgusted at this simple mistake, the temporary laborers were dismissed but still had to be paid for four hours of work although they were there only for a maximum of thirty minutes.

I spent the rest of the weekend relaying out the warehouse on my own time, despite the fact that other plans with his wife had been made. After making an overall outline of the warehouse, the first thing I did was to permanently locate the posts and highlighted them in bold. So why did I commit this mistake and what could I have done differently? I wanted to demonstrate some of the knowledge gained during the various courses taken in college. The course that was not taken was the one in critical thinking; what questions should be asked and to whom?

This mistake was definitely avoidable but occurred because I had not taken all aspects of the project into consideration. I did not consider the obvious which was something that was difficult not to observe. To make matters even worse, there were many such posts that were hard to avoid since they were painted bright orange. How could I have missed them? Regardless of the project with which I am now involved, the first thing I do is to locate and number the vertical uprights and on the drawing. I also include horizontal obstacles to make sure that a forklift does not unnecessarily tear the building down as a truck is being loaded. I tried painting the vertical uprights traffic yellow but even that did not work, so now I hang a tennis ball as a reminder of its location just as some people have a tennis ball hanging in the garage to remind them of how far to drive to prevent driving into the wall. I even use a multicolored tennis ball, because it attracts more attention and is cheaper.

I had a job with a paper converter with a sister company that was a paper mill manufacturing paper from recycled material. The mill received newsprint in railcars but also used cartons that were brought in daily from local suppliers usually in individual pickup trucks. The process was that the pickup truck would enter the mill's yard, weigh in on scales, drive around to the rear of the mill, unload the corrugated waste, return to the scales for a reweighing, and receive in cash the value of the waste based on the current value of corrugated scrap.

One day, I was walking with the chairman of the mill and told him that I was going to retire and that the chairman had to fund his retirement. The chairman looked flabbergasted and had that "deer in the headlight" look. I then explained how the chairman would fund his retirement legally.

First, I would purchase two identical trucks and park them a block away from the mill. I would then fill one truck with corrugated waste, drive to the mill, get weighed, drive around back but instead of unloading continue driving until I was next to the empty truck, swap tags or license plates and drive across the scales with the empty truck, get weighed and receive cash for essentially nothing except my time. I would then drive the empty truck next to the truck with the bed full of corrugated waste, swap tags, drive to the mill, get weighed, drive around back, but instead of unloading continue driving until I was next to the empty truck, swap tags and drive across the scales with the empty truck and get paid in cash for essentially nothing except my time. I would continue the process until I had enough cash to deposit that day to avoid suspicion.

The chairman looked at me in disbelief stating that no one would ever take advantage of the company in this manner. Employees here were dedicated and devoted to the company for the long term. Within several weeks, the chairman and the president of the company for which I worked came into my office and presented me with an opportunity. The actual cost to produce paper board had been increasing over the last six months although there had not been a cost increase in any of the ingredients, particularly corrugated scrap which was over half of the cost of manufactured board. The chairman had started thinking seriously about someone using my retirement idea and was curious as to whether if it was being employed.

My boss asked me to investigate the situation and report the results. I would randomly check trucks about a quarter mile from the weigh station and discovered that the person at the weigh station was skimming off the top. The receipt was correct but the amount of change returned was less than correct and because most people do not check their receipt against their change, the difference was not noticed. If the difference was noticed, the person in the booth just explained that they were sorry for the error, apologized, and provided the difference.

This mistake resulted primarily from the failure to consider all aspects of the operation. Everyone assumed that the correct change was given as each vehicle exited the weigh station. Extra precautionary steps must be built into a system to prevent such flagrant cheating. The person who operated the waste house was a trusted long-term employee. The quantity of money siphoned off over a period of six months exceeded many thousands of dollars, so this mistake was not nickel and dime.

To quote a favorite proverb of Vladimir Lenin, *"doveryai, no proveryai"*, which means "trust, but verify" made famous by President Ronald Reagan. This phrase was repeated many times by President Reagan during various meetings with the Russian leader, Mikhail Gorbachev [16].

Another mistake was due to the lack of involving all aspects of the project, the failure to ensure that a product was fully able to be manufactured before it was illustrated in a catalog with prices so that orders could be taken. Because the company had been manufacturing desktops for many years, it assumed that it possessed the knowledge, skill, and manufacturing capability of producing this particular desktop to meet the quality and quantity specifications required. This product was a new design that involved a metal insert that was to be placed on the tabletop.

The insert was to be placed in a groove cut into an elliptical shape into the desktop, requiring the precision that is achievable with a CNC (computer numerically controlled) router or similar machine. A CNC machine can have as many as five axes and is a motorized maneuverable tool controlled by a computer program to perform certain operations that are programmed into the software.

The situation was that the company had a CNC machine that was not capable of accepting updated software due to its age. Attempts were made to produce a fixture that would allow manufacturing to route the curve manually but after numerous efforts with several different operators, the management recognized that further efforts would be futile. Routing of the top occurred after the veneer had been cut and edge glued into a face, which is several inches larger in each dimension to accommodate further processing. The face is then glued onto a piece of medium density fiberboard (MDF) board on one side and a sheet of paperboard on the bottom. The assembly is then edge banded and the final operation is routing. Thus, if a desktop is scrapped at this step in the manufacturing operation, then it has a large percentage of its material and direct labor costs (Figures 4.1 and 4.2).

My first experience with a furniture company was with a small one located in the middle of nowhere, North Carolina (NC). It produced a variety of low-end dinette sets consisting of tables and chairs. The chairs could be purchased in a number of different combinations of fabrics and vinyl. The tables were produced in a variety of styles in several laminates. Prototype products of new styles were actually built to ensure that the products could be produced and to eliminate any production problems before they were shown in the semi-annual Furniture Market in High Point, NC. For every product shown, an accurate bill of material and routing was determined prior to be shown and entered into a computer program that determined the actual cost. For any product, anyone in the company was able to visualize on the computer or print the BOM, the actual cost, the vendor, the purchased cost, the routing of the product, the sales history, the current inventory, and demand history. For those products that were shown at the market, anyone in the company was able to visualize or print everything except the inventory and demand history since none existed. Before any product was introduced at the market, the standard direct labor and material cost, the processes involved in producing the product, the vendors, order quantity, delivery terms were known, etc., because if an order from a vendor at the semi-annual market were to be received, the company must be able to deliver the

TITLE	Desk top with Inlay (orthographic view)			
Figure Number	4.1 Drawn By	JD	Date	7/25/2019
Desk top with Inlay				
Checked by JW	Date	8/15/2019 ECO #	978-002	

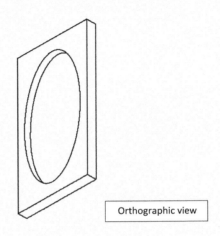

Orthographic view

FIGURE 4.1   Desktop with inlay—orthogonal view.

product profitably when desired and in the quantity required. The vendor may order any quantity from one to a truckload with a delivery date from one to several weeks.

For an individual who had worked for a much smaller company that knew all vendors, the BOM, the routing, and the costs of a product before it was marketed, this situation was inconceivable for any company much less one of this size. This company was much larger when measured in sales, number of employees, number of plants, etc., but the product was advertised and priced without the assured knowledge of manufacturability. This would have never occurred at the smaller company for which I had worked. I was not sure how the larger company handled this situation but I was sure that I would not want to admit to a large customer that the company sold a product that it had not produced a prototype before marketing it.

One project I inherited while working as a project chief of a government facility was the renovation of a building to enable some employees to be relocated from several areas of the campus to this building. Of course, the first question I asked was if these employees had been working in their existing locations for all these years and why did they need to be relocated? And the second, what was the cost justification for the move? Apparently, there are some questions that are just not supposed to be asked. The project had already been approved so the answers were moot and never received. Since the building into which the employees were to be moved was on the historic register and had not been occupied for several years, the restoration required

TITLE	Desk top with Inlay (plan, top, and right-hand side views)		
Figure Number	4.2 Drawn By JD	Date	7/25/2019
Desk top with Inlay			
Checked by JW	Date	8/15/2019 ECO #	978-002

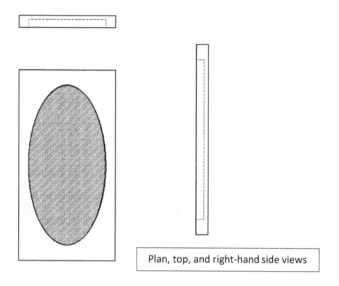

Plan, top, and right-hand side views

FIGURE 4.2   Desktop with inlay—top, plan, and right-hand side views.

additional inspection procedures to pledge that it was returned to its original or better shape. One example was the slate roof. Another was asbestos removal and remediation which is an expensive and lengthy process.

One feature of this building was a chiller that moved heat from one location to another. The original chiller was wired in 220 V and the project engineer did not take the ten minutes needed to drive in a government-supplied vehicle to this building from his office to retrieve the information from the data plate. This is Engineering 101; obtain accurate information from the existing equipment if you are going to replace existing equipment with the same type. For some reason, he assumed that the chiller was 440 V, the same voltage as everything else in that building and on the facility.

The chiller arrived and on installation, the contractor realized that there was a problem. Replacing the electric motor was not possible due to time constraints. Hence, a converter had to be ordered. Shipping added a delay plus the contractor added an additional charge for the installation of the converter. This mistake cost many thousands of dollars and was easily avoidable if the project engineer had looked at the data plate on the old chiller.

People began to relocate in April into the old building and since the transformer was ordered at a later date with a delivery date in November it would require that everyone suffer through the summer in the old building. To avoid that, everyone was moved back to their original offices at an additional cost only to be relocated again at an additional cost once the air conditioning became operational in November when it was no longer needed. The heating component of the chiller worked fine.

The engineering technician failed to consider all aspects of the project. Not only was the data on the chiller overlooked, but he failed to take into the account what effect the temperature change and people traffic would have on newly installed inexpensive vinyl floors. The tile floor buckled excessively, which caused a tripping hazard requiring that the floors be replaced. The contractor refused to replace the floor under warranty due to the circumstances under which the floor was installed. The second time the floor was replaced with an upgraded tile and adhesive that was guaranteed not to buckle. The additional cost was incidental compared to the costs of replacement.

This mistake was very costly and was entirely avoidable. It resulted from the failure to consider all factors involved in the project. Again the use of critical thinking would have caused these possibilities to have been realized. Failure to retrieve proper information from the data plate from the chiller and to determine the effects of temperature differences and the human traffic on vinyl tile were the leading causes.

In one furniture company in which I worked, a work sampling study was conducted to determine the number of material handlers needed in the receiving department. Material handlers were responsible for unloading, verifying the quantity against the bill of lading, scanning the bill of lading to receive the shipment into raw material inventory, notifying the purchasing of any discrepancies, and finally storing the material in the proper location(s).

A work sampling study is a statistical study performed to obtain information as to what percentage a person or group of people are working or not working. The only objective of the study is to determine if a person is at his designated work station and is working or not working. It is a binary study—either he is at his designated work station or not, and if there, whether he is working or not. Working is defined by performing one of the tasks in his job description. If he is not there, then you are not concerned as to his location—the only known fact is that he is not at the worksite and therefore cannot be working. One begins the study with a set of random numbers converted into times and eliminating those during and close to established break and meal times to prevent bias. Results must be repeated over several different days to be meaningful. Results of a work sampling study are not intended to be as precise as those of a time study but are intended to provide guidelines only. For example, if the results reveal that you need four and you have three or five then you are probably okay. If the results indicate that you need four and you have ten, you can assume you are overstaffed and you should look for other tasks that the crew can perform. After reassigning a crew member, perform another work sampling and proceed as the results indicate.

Before applying the results, one must think of the possible consequences of the effect that bring a change in the size of the receiving crew. An increase in the size will result in a reduction in the amount of time required to unload a truck and store its contents into the proper inventory locations, whereas a decrease would have the opposite effect. Results of the work sampling study revealed that the receiving crew was overstaffed by several members.

A similar study was conducted but this time the focus was to determine the material that consumed the most manpower to receive, unload, and store in its proper location. The product was foam. This was corroborated by an analysis of inventory receipts that revealed that foam was not only one of the most frequent items ordered, but a review of the volume consumed in the warehouse indicated that foam was also the bulkiest item that consumed much of the warehouse space.

Returning to the results of the work sampling study to determine the percentage of the time the material handlers spent in placing the foam into proper locations, the study team realized that material handlers could take the foam to the production line if the truck was loaded in the proper sequence in less time than placing the foam in proper locations. Taking the foam to the line instead of to the proper locations would eliminate the subsequent action of taking the foam from the proper location to the production line. The foam was stored in racks that were three positions high requiring the use of a ladder for removal. The company was very safety conscious so that the use of a ladder required two people, one to hold the ladder while the other was on it.

The delivery of the foam was to occur one hour before it was actually needed, which would provide the material handlers sufficient time to perform their receiving duties before taking the foam to the production stations. This proposed change was discussed with the management before proceeding and engineering received their agreement to proceed. Purchasing met with the production control to determine the order that the truck should be loaded from the rear, defined as the location of the loading doors. Then the team met with all involved to make them aware of the changes about to occur—that JIT was actually was going to happen even if the plant was in the middle of nowhere.

Everything went smoothly for the first six months. Then one day the truck delivering the foam did not appear at the designated time, 6:30 am and there was no phone call. This was before cell phones. The foam vendor was approximately 45 minutes away and called at 8:30 am informing the plant that the truck had run out of gas en route and would be at the plant in approximately an hour. That truck driver had been making that delivery forever and never had a problem. The fact that the truck was late and the plant operated on a JIT basis translated into having to shut the plant down since departments downstream did not have any parts to manufacture and those upstream could produce parts but there was no place to send them for further processing. The result was that the entire plant was shut down.

This costly mistake would not have occurred, if critical thinking had happened and all aspects of the project had been considered. But no one ever thought about the delivery truck having a mechanical problem or running out of gas, or the effect that not having foam would have on the operations on the plant. Unfortunately, both the company and the vendor gained valuable information from that situation. The lesson that the company learned was to have foam in sufficient quantity to manufacture stock items for two hours and for the vendor to always check gas supply before he would leave and to have a second tractor readily available to haul the trailer.

On a sales call to a potential customer, a packaging engineer asked if the company could supply what he termed a core plug (Figures 4.3 and 4.4).

It was designed to secure the stretch film used to package the final product by fitting after the stretch film was applied with a ½ inch lip to prevent slippage inside the paper tube before the placement of a shrink bag over the entire pallet which was then shrunk to ensure package integrity.

The core plug consisted of a paper tube that a diameter small enough to fit inside the tube onto which the product made by the company and had a larger tube stapled at one end that acted as a stop. The thickness of both tubes was measured from the existing vendor and found to be 0.500 inches. The tube acting as the stop was stapled onto the other tube.

My previous employer was at a facility working with mentally and physically handicapped individuals and thought that this would be a perfect opportunity for them to learn a new skill and earn money. The process of stapling required that the tube to be stapled first had to be cut with a bandsaw to facilitate insertion onto the smaller tube, then handed to an employee who activated a stapler operated by a foot pedal, stacked the assembly into a corrugated tray, and then packaged the unit using stretch film. New skills would be learned by clients as well as it provided additional opportunities to earn money for the client as well as the facility.

With my previous employer, I was the plant manager and acted accordingly. Shipment deadlines and quality standards were to be met or exceeded at all costs.

TITLE	Core Plug (orthographic view)			
Figure Number	4.3 Drawn By JD		Date	8/10/2019
Core Plug				
Checked by JW	Date	8/20/2019 ECO #	978-003	

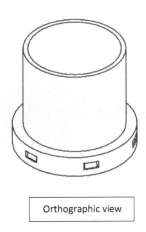

Orthographic view

FIGURE 4.3 Core plug—orthogonal view.

TITLE	Core Plug (plan, top, and right-hand side views)		
Figure Number	4.4 Drawn By JD	Date	8/10/2019
Core Plug			
Checked by JW	Date 8/20/2019 ECO #	978-003	

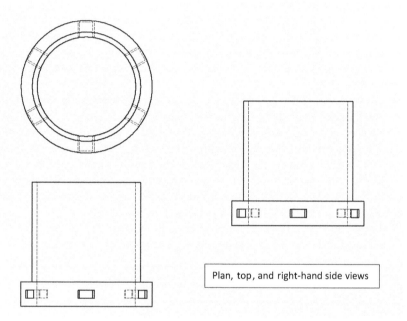

Plan, top, and right-hand side views

FIGURE 4.4 Core plug—top, plan, and right-hand side views.

He established minimum production standards and set goals with rewards for clients who achieved those goals. Unfortunately, the predecessor did not have the same goals. He talked with the plant manager about the feasibility of the contract and the weekly requirements. He agreed that there would be no quality- or quantity-related problems. Shipments of both tubes and supplies needed were made to the facility.

Several days later, I called the facility and spoke with the plant manager to ask about the progress and possible pickup of finished products for delivery that following Monday. He responded that none had been produced due to the lack of available clients. I kindly reminded him of his commitment that they both looked at each other in the eyes and shook hands while agreeing and that the delivery must be made. His response was not very positive.

I then spoke with the plant manager where I worked and explained the situation and requested that I take three of his plant employees to the facility the next morning to work on these core plugs as long as needed so that pickup could occur Friday for delivery the following Monday. The manager of the plant agreed that it was the only viable solution at that point. I also explained to him that the next time that there was a problem with the facility not being able to deliver as promised then all production would be relocated to his plant.

I called the plant manager of the facility the following Tuesday to inquire about progress. Again I was told that nothing had been done and that if I was going to continue to bother him about production, then he would prefer not to have the contract. I responded with a statement to the effect that this could be easily arranged.

Again I spoke to his plant manager but this time suggested that instead of just sending three people to work on the core plugs, the company send a truck and crew to return their supplies, and set up an area in the plant in which I worked to produce the core plugs to make sure that delivery times were met.

The mistake was definitely avoidable, had I employed critical thinking and did not assume that his replacement shared the same goals for clients and organizations as mine concerning increasing opportunities for clients to learn new skills and earn more money by attracting new and different contracts.

5 Failure to Ask Others for Assistance

Another mistake that I witnessed was almost fatal. A contractor had been hired to refurbish a room in a hospital, requiring the removal and replacement of electrical fixtures and outlets, removal and replacement of walls and doors, and general construction. Part of the contract was that this federal government agency would provide electricians required within a ten-minute notice during the construction to warrant that electrical safety was maintained and that all work met the existing code. The person performing the work for the contractor was paid by the hour, so he was paid for the amount of wait time he incurred.

During a meeting, I received an emergency message concerning the area in which this reconstruction was occurring. On entering the area, I immediately noticed that electric lights were not working, which indicated a problem. As I entered the area under renovation, I observed several obvious electrical violations. First, all wires were just hanging down with no wire nuts or electrical tape covering them and second, the contractor was using an aluminum ladder. Apparently, the contractor, while relocating the ladder had touched a live wire, tripping a switch that controlled, among other things, the operating room. There was a patient on the operating table, but the operating room lost the lights before the operation began. The surgeon and I had an unpleasant chat. Previous discussions with colonels were a walk in the park compared with the one with the surgeon.

After the chaos was normalized, I called the contractor's company and explained the situation in detail. I indicated to the contractor's company that I had the police to escort his employee off the premises and that the employee was not to return under any circumstances. I then wrote an incident report, so that everyone would be aware of what had occurred and knew the cause for this event. I also recommended to the safety department that aluminum ladders be banned in future from this government facility, if not all government facilities.

This mistake could have been definitely avoided. First, the contractor was told to advise the management when an electrician would be needed so that proper steps could be taken. The contractor was working by the hour, so his pay would not be affected by an increase in the time required to complete the project. Second, an aluminum ladder should never be used in the vicinity of electricity since aluminum is a good conductor of electricity. Due to the shortage and high price of copper and greater availability of aluminum, companies began the development of copper-cladded aluminum (CCA), with rolled sheets being the first product. Wire was developed at a later date. Due to lower costs and availability, CCA wire was used and thought to be the answer to the residential housing situation, but it failed miserably. A survey conducted by the Consumer Product Safety Commission concluded that homes built before 1972 and wired with aluminum had a probability of 55 times greater to have a connection attain Fire Hazard Condition than homes wired with copper [17].

This mistake occurred for a number of reasons as has been shown, but the overriding one was the failure to ask for an assistance. In this case, the assistance would have been willingly provided. The electricians would have never used an aluminum ladder and they would have been available for support as needed. Most importantly for the contractor, the support from the trained electricians would have been provided at no cost.

Third, this employee was working by himself—he did not have a companion employee in case of an emergency situation. This potential problem would have been eliminated by the requirement that on all future contracts, a minimum of two employees were required to be working on-site at all times.

One company for which I worked produced office furniture made primarily of wood. The first process in the manufacture began in the mill room where dried purchased lumber was machined via numerous processes for assembly into various components that ultimately became saleable units. For all components, the last routing operation included a sanding attachment to potentially eliminate the need for a subsequent time-consuming manual sanding operation. For example, the final operation in the machine room for chair arms, lounge chairs, and sofas was a profiling operation that contained a sanding attachment that sanded all four sides of the arms prior to the arms being discharged. Initially, the grit of sandpaper used was 150 but was changed to 180 without engineering or management's approval.

Productivity in the Hand Sanding Department was being monitored for several reasons. First, everything had to go through this department and the quality of the output was critical to the quality of the finish that was applied. It was the bottleneck operation in the plant. This was the motivating factor for the addition of the sanding attachments to some woodworking machines, because these attachments should reduce the workload and therefore increase productivity. However, the productivity was on a downward trend.

In an attempt to discover the causes for this discrepancy, it was decided to hold a Kaizen event. This term is translated into English as "continuous improvement". Although the concept can be traced to Dr. Edward Deming, it was the quality guru, Masaki Imai in Japan, who made the concept popular worldwide [18]. The term literally is translated as Kai (change) and Zen (for the better.

The supervisors of both departments and engineers gathered at the Hand Sanding Department anxious to solve this problem. The group decided to study several employees, take notes, and then proceed to the next employee. Within ten minutes, the group realized that the employee in the Hand Sanding Department began sanding the chair with 150 grit sandpaper. These are the components that were sanded in the machine room by the attachments with 180 grit sandpaper. The problem with using a lower grit sandpaper on subsequent sanding operations is that it will introduce scratches and dents that must now be removed by additional sanding, usually with a higher grit sandpaper (180). To state the problem simply, the use of 180 grit sandpaper in the machine room caused more work in the Hand Sanding Department.

The solution was simply to change the grit on the sanding attachments in the machine room to 150. Conversion to 180 sandpaper in Hand Sanding was not feasible, because all components were not sanded in the machine room [19]. Consistency and standardization were critical for training and process sheets. This mistake was costly and definitely avoidable. It occurred because the initiator had not considered all variables involved. There was a process in place, the Engineering Change Order (ECO) system, to make changes, and it was not followed. The ECO system is a formal method of asking others for assistance.

I worked for a large company that manufactured labels that were printed on narrow fabric with the primary plant in the middle of nowhere in the Southeastern United States. This plant produced printed labels that were attached to underwear and outerwear to identify the producer of the garment in addition to providing washing instructions for the garment. The company also had a facility in New England that produced labels using a Jacquard loom. The Jacquard loom was invented by Joseph Marie Jacquard who gave his first demonstration in 1801 and received a patent in 1804, and was recognized as a revolution in human–machine interaction due to its use of the binary code, either there was a punched hole or not in a punch card in a column. This ability to interchange punch cards was the inspiration for the design of the early computer.

The company recently had purchased an off-the-shelf computer system that required numerous modifications. The staff at the plant at which I worked were the most computer-literate and had the most manufacturing experience of any of the other plants. These two facts facilitated the implementation of the computer system. There were several staff members at this plant with previous experience with integrated manufacturing computer systems which provided the staff with the knowledge to enable them to ask pertinent questions when visited by the system's consultants. Implementation at my plant was not transparent but given that the purchased system required many modifications, implementation was rather smooth.

This was not the situation in the New England plant. The personnel failed to ask necessary questions of the system's consultants, primarily because they did not know what questions to ask. No one at the plant performed critical thinking to determine who should be asked what questions? The situation reached a crisis when the chairman of the company called the general manager of the plant in which I worked and told him to send a team to the plant in New Jersey, because one of their biggest customers was about to cancel all of their orders due to a non-shipment of any product for a week. This coincided with the time the new computer system was installed.

The general manager at the plant where I worked called the manager at the New England plant for corroboration since failure to ship products to a customer, particularly a large one, was not an option. When the plant manager confirmed the story, the general manager informed him that a team would be there tomorrow to address his problem. When asked why no product had been shipped, the plant manager informed the general manager that no one was able to get the printer to print shipping labels. At that point, the general manager told the plant manager to have his shipping crew use the same thing that he had used to pick up and dial the telephone—his hand. The plant manager then said that the shipping ticket was tied to the invoicing and he was not sure how the customer would get invoiced. It was at that point that I, who was witnessing this not so pleasant conversation, thought the general manager was going to explode, but he calmly stated that if necessary this was an early Christmas present.

It would be better in the long run for all concerned to simply give the company the labels rather than take the chance of losing this large or any customer over a few hundred dollars. He then stated that the customer always must come first.

This mistake was avoidable and occurred solely due to the failure to ask for help from either the consultants or from the staff of the plant at which I worked. The mistake could have resulted in the loss of many customers and jobs.

Another preventable and costly mistake occurred after a sales call I made with a salesman to a plant that manufactured paper used in copy machines. It is a large company with manufacturing plants nationwide. The purchasing agent provided the salesman with a detailed specifications sheet, which the salesman quickly looked at and responded that he had no issues. The sales call was made on a Tuesday; the purchasing agent stated that they would need a truckload of tubes the following Monday. Our salesman responded with, what time should the truck be at the customer's location?

After examining the specifications, I realized these tubes would need to be produced in another plant due to the additional processes required. The additional freight or the cost of the additional processes had not been added into the price quoted to the PA. I did not question the salesman, as doing so at that time would not have provided an organized company. Those questions should have been addressed and resolved had the specifications been made available to the salesman and the engineer prior to the appearance at the PA's office. However, given the situation, if the PA had asked me, I would have suggested that the company supply a 1M test samples for a test run to be monitored by me, the company's engineer, and QA, to determine how well the company's product performed and make recommendations as to the next step.

The truckload of tubes was shipped and received that Monday. I received a call from the PA within several hours of receipt (the receiving company was in a different time zone), strongly requesting that I and the salesman be in his office at 8 am the following morning if my company desired a chance to salvage the already blown opportunity to conduct business. As I hung up the phone, I thought the meeting would probably not be a pleasant one.

When I arrived at the PA's office, the PA indicated that he had received a call from his boss before he called me. The conversation with his boss was not very congenial. He then took me and the salesman onto the plant floor. He stated that the tubes my company shipped varied in length and outside diameter (OD) and explained in detail the problems that it caused to the production in the plant. Due to the speed and width of the paper roll run, tube length and OD variations resulted in machine stoppages and cost overruns, which more than offset the few pennies that the company was going to save by changing tube vendors.

I and the salesman then returned to the PA's office. He looked at the salesman and asked if he read the specifications, to which the salesman said "No." Then the PA asked why he shipped what he did. The salesman said that his company supplied the PA's plant in Dallas, TX, the plant was also making copy paper and that the specifications for that plant were only two pages. The PA quickly stated that there was a reason that their plant had much more detailed specifications and that if the salesman had taken the time to read them he would have found out what the difference was.

The PA also stated that this plant was not in Dallas. At that time, I thought the PA was going to scream but instead he calmly asked him what he would have done after having examined the specifications. I responded that he would offer a sample of 1,000 (M) tubes to be monitored for results by him, the company's engineer, and QA, to determine how well the product performed and make recommendations as to the next step.

The PA then calmly called the president of my company and told him in no uncertain terms that he never wanted to see that salesman again. The only person he wanted to see was me. After hanging up the phone, the PA asked me if his company wanted to be a vendor to which I responded positively. He then asked how soon the samples could be shipped. I responded that at this time it was not known since the company must first ensure that it could guarantee that it could manufacture the product to meet or exceed the customer's specifications. The PA would be advised of the progress. Within a month, the company became an excellent customer.

This mistake resulted from the failure of the salesman to ask for the assistance of the QA lab. The salesman assumed that since the company was servicing one plant all company plants would use the same product. First, the salesman should have obtained several tubes from the customer to verify and check against the specifications he was provided. Then, he should have determined which plant was able to consistently produce tubes that met or exceeded the tubes that he received from the customer. Then and only then can he comfortably quote prices for the customer.

The question that must be asked is what if there is a difference between the specifications provided by the vendor and those determined from the samples received from the customer.

First, everyone knows about samples and populations. As one who has incurred this exact situation numerous times, I had gone back to the customer and asked for a few additional samples for testing. One always speaks highly of competitors. Regardless of the outcome, I had always provided the customer with a product that met or exceeded the provided specifications. Only after a potential customer becomes an actual one should testing begin with the customers' approval to possibly reduce specifications and costs.

One aspect of the job that I enjoyed was traveling with salesmen to meet customers to assist in solving problems or providing technical assistance. The company served a large manufacturer of vinyl-backed wallpaper. The current method of packaging was all manual and consisted of several steps. After the roll was produced and wound onto the paper tube, the roll was placed onto a conveyor belt where a team wrapped the roll in brown paper, secured it with masking tape, and placed the wrapped roll onto a pallet. The rolls were then manually placed onto a pallet where each additional layer was placed perpendicular to the layer below. This arrangement is usually referred to as the tubes were "pig-penned" onto the pallet for shipment (Figures 5.1 and 5.2).

The pallets were then wrapped with plastic straps and then strapped onto the pallet using polypropylene straps, which is safer and less expensive than steel straps but more expensive than plastic strapping that had a tendency to stretch.

The plant engineer explained that the current method of packing was labor-intensive and therefore expensive, and his proposed method was to be implemented the

TITLE	Pallet illustrating "pig-penned" paper tubes (orthographic view)				
Figure Number		5.1 Drawn By JD		Date	8/25/2019
Pallet illustrating "pig-penned" paper paper tubes					
Checked by JW	Date		9/2/2019 ECO #	978-004	

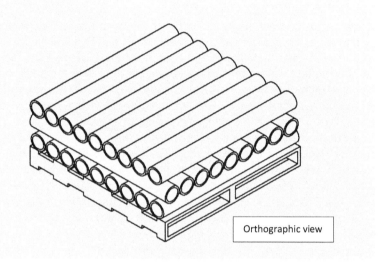

Orthographic view

FIGURE 5.1   Pallet illustrating "pig-penned" tubes—orthogonal view.

following Monday after the equipment had been installed over the weekend. This new packaging method consisted of automatically inserting the roll of wallpaper using robots into a plastic bag for shrinking, placing the shrunk roll of wallpaper onto a pallet, then automatically placing a shrink bag over the entire pallet, conveying the pallet into a shrink tunnel where heat was applied to shrink the bag, and then ejecting the pallet onto the plant floor to enable a forklift to load the pallet onto a trailer. He advised the calculated payback was less than six months. The plant engineer's only concern was the effect of the heat on the paper tubes from the shrinking of the bags.

My concern focused on whether the heat would result in causing the vinyl to fuse together with the wallpaper. When the plant engineer said that he had asked the same with the engineers of the companies that manufactured the heat shrinking equipment and was assured that the heat would have no effect, then I really began to get concerned for a fellow engineer. I asked him if he had tried this process out personally before recommending the change and he responded that of course he had not. I then asked if there was a hairdryer around that could be used for reassurance that this process would work as planned. He responded that he did not have time since the equipment was on site, work orders had been signed to installed it that weekend, crews had been trained, and customers had been advised. The only question that was not asked was if he had time to look for another job before the weekend.

TITLE	Pallet illustrating "pig-penned" tubes (plan, top & righthand side views)			
Figure Number	5.2 Drawn By	JD	Date	8/25/2019
Pallet illustrating "pig-penned" paper paper tubes				
Checked by JW	Date	9/2/2019 ECO #	978-004	

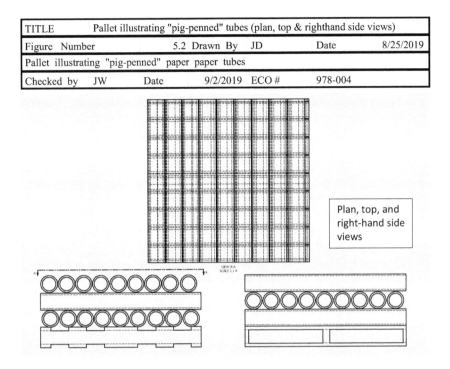

Plan, top, and right-hand side views

FIGURE 5.2   Pallet illustrating "pig-penned" tubes—top, plan, and right-hand side views.

The following Monday, I called to talk with the engineer over the project only to discover he was no longer with the company. This mistake cost this engineer his job, his reputation, and not to mention the many dollars the company lost as a result of the lack of critical thinking. His desire to save money for the company and improve processes was the purpose of his position, but he allowed the desire to demonstrate results quickly to overcome proven tried and true methods for process improvement. First, ensure that the process works for your product; second, check to determine if other companies that produce similar products use this process; three, document both your successes and failures and its reasons; and four, proceed in a logical and methodical manner to ensure that the results are meaningful and reproducible.

This mistake was definitely avoidable and occurred due to a number of reasons. However, the primary one was the failure to ask for assistance in this project. The engineer should have talked with and visited other companies that use this method of packaging to discuss success and potential problems. Use of critical thinking would have suggested the feasibility of the adhesion between the plastic-backed wallpaper and the plastic packaging. Then, this possibility could have been tested with something as simple as a hairdryer. Failure to determine if there would be adhesion between vinyl-backed wallpaper and plastic film resulted in the unnecessary expenditure of thousands of dollars spent on equipment and labor on installation and the loss of one job, all of which could have been prevented with simply testing using a hairdryer.

For ten months, I was chief of the projects section of a government facility that had two engineering technicians and an interior decorator who reported to me. The first week of employment was a week of orientation, a meet-and-greet week, and orientations to various departments within this government facility, the police, HR, housekeeping, etc. During the director's introductory speech, he mentioned that although the number of people that this government facility was servicing was decreasing, the number of employees at the government facility was increasing. As an industrial engineer (IE), my experience was that the opposite should occur. So when he asked if there were any questions, I just had to ask the question as to why those numbers were positively related instead of inversely related since the employees were there to serve the people. The second mistake I made that day was providing the name when he asked but that was unnecessary since it was on my nametag.

One of the projects that were being worked on was a healing garden for the people who lived in the facility. The engineering technician requested that the color to be green, and the contracting officer did not object to the color being green or request further clarification. On arriving at the job, the first question I asked was what color of green. The response was the color of grass. I then asked what kind of grass. Well, since I was the new guy, all of a sudden I became a bad guy because I was the one asking all these questions. The new boss told me that both the engineering technician and the contracting officer were upset, and it was my fault just because of my desire to be specific. The maintenance facility contained a Pantone Matching System (PMS) color chart. All that was required was that the engineering technician consult with the interior decorator, agree on a PMS color, and write that into the specifications. The engineering technician was available at most a five-minute walk from the PMS chart that was located in the same building in which he had an office.

Because the color—other than green—was not specified in the contract, the contractor had a legal right to enter a change order for the contract although nothing had changed. The only thing that changed was that the generic color green had been specified to a certain PMS color. The contract was delayed as a result to enable the contractor to write and get the additional paperwork for additional monies approved.

This mistake was avoidable and did not have to occur. Unfortunately, due to a lack of critical thinking, it was not the only costly preventable mistake that occurred with this contract.

The healing garden was to also have electric wires installed to melt any ice and snow that accumulated to prevent any person from falling. Again this project had already been approved for funding. When I asked the engineering technician the maximum amount of electricity that would be required on the coldest day with the most snow, the response was that they did not know and had performed no calculations for the least or most amount needed. The government facility was privileged to have a senior electrician who was capable of providing an answer within a few minutes to those questions if asked. This senior electrician was in the adjacent office. Walking to this senior electrician would have taken at most five minutes.

The senior electrician was not asked or contacted in any manner. When the correct solution was calculated by this senior electrician who provided a solution that appeared initially impossible, the local provider of electricity was called who sent an electrical engineer who confirmed the solution. The solution was the maximum number of ampere needed for the healing garden exceeded the suppliers' ability to supply the entire town. This changed the entire scope of the project. The healing garden had to be withdrawn from the project, due to the failure to consider the amount of electricity that would be needed. But from then on, I was known as the person who took away the peoples' healing garden. Needless to say, not many Christmas cards or gifts that year were received by me.

These mistakes were easily preventable if the technician had asked others more knowledgeable than him to obtain additional information needed to proceed with the project. One cannot just select a color from the rainbow—one must be specific. One cannot state that a project will use just a little bit of electricity—how one determines the cost of a little bit. This is similar to the time when I asked my boss that I would like to perform a bigger check forgetting that a sister company was the one that produced paper board. The next check I received had to roll up on a paper tube to be taken to the bank but it was a bigger check. I reminded my boss that the intent was to have a check with digits that were more to the right of zero on the number line but I did receive a bigger check. I was sure that my students were tired of telling them that the instructor was anally-retented but one must ensure that there was no misunderstanding. That is the reason why if there is a quantity with a zero before the decimal point as 0.6 pts, then I will always write 0.6 pts.

While conducting the research for my dissertation, I discovered that the numerous medication errors resulted from the absence of a leading zero in the writing of medications, which resulted in a council recommendation to enhance the accuracy of prescription writing revision, dated June 2, 2005 [20]. Never let the reader or anyone assume anything. As previously mentioned, I assumed that when asked for a bigger check his boss would issue one worth more money. But I did receive a bigger check but one that had the same monetary value. I learned from the situation and used it as a teaching aid. Always include units—there is a difference between pts and gallons. I make sure that my students know the difference when offered a job—there is a difference between $50,000 a month and $50,000 a year.

Reasons for these mistakes are simply that the employee fails to ask and take the advice of experts. This project was the one that should have had the input of several people initially. The facility had employees with many years of experience in different areas. A good project leader always takes advantage of the experience and education of those around him for several reasons; one, it adds to their knowledge base; two, it makes them a part of the team; and three, it increases the likelihood of a successful project.

During the first quarter at college, I was required to take Reserve Officer Training Corps (ROTC) since my college is a land grant school. In the first quarter, I was introduced to the phonetic alphabet. Initially, I thought, oh great, something else that must be remembered. But when talking with people whose English is not their primary language or when one is under stress, it is easy to interpret "a, b" as "a, d" or "a, c" as "a, z." There are many letters that sound alike, especially when one talks fast or talks with one whose command of the English language is not as good as the native speaker's. Now, I had that person repeat the message using the phonetic alphabet—it might require more time to convey the message but the number of mistakes in communication was eliminated. Elimination of mistakes is more important than saving time, since in the long run the elimination of mistakes will save time and money.

As the industrial engineering manager of a company that manufactured paper tubes for the carpet and textile industry, I was advised that a small company was producing a better product at a lower price. The company, for whom I worked, hired a consultant, a chemical engineer, to determine the process the smaller company was using. The process of producing the paper tube—think of straw—is first extracting from a roll of paperboard a certain length, the length of the tube, applying adhesive on the inside, then rolling the board with the applied adhesive around a mandrel with a horizontal slit that accepts one edge of the paper, the diameter of which was the diameter of the tube, followed by drying, then bundled into bundles of 37 or 51, depending on customer specifications, using a fixture and strapping via steel strap, and loading into a trailer with a forklift for shipping. Almost half of the cost was the paper board. The strength of the tube was end to end and resulted from the fibers from the board which ran parallel with the length. This strength is called beam strength and is the reason this type of tube was appropriate for this application.

This smaller competitor was replacing a pound of paperboard with a pound of marble dust. Marble dust is finely crushed granite rock that can also be used as an alternative to sand when constructing a driveway or walkway. The weight of the tube is the same but marble dust costs much less than the paperboard that it replaced. It was important that tube weight did not change due to tare weights. The marble dust when mixed with the adhesive becomes very hard and brittle when it dries, thus producing a tube with a much higher beam strength at a much lower cost. At the time, the cost of marble dust was approximately 20 percent the price of paperboard, so its use resulted in significant improvements in profitability.

My boss decided to take advantage of this and follow the same procedure and based on the formulation provided by the consultant, a process was developed and marble dust was ordered. The process was implemented and the company began saving money using the new process. Everything was going well until the company was notified that it was being sued for patent infringement. The company was in such a hurry to save money and was confident of its ability to proceed. Due to its size relative to the smaller one, it had failed to hire an attorney to conduct a patent search.

This was a huge, avoidable, and costly mistake. After months of threatened litigation, the company purchased the smaller one at a premium price. In addition, my company had a septic tank in one plant for a sewage system, whereas the acquired company used the city sewer system. As soon as the company began adding marble dust to the adhesive, no new adjustments were made to the septic tank system resulting in some marble dust being released into a stream causing the death of farm animals. So, this mistake resulted in not only the purchase of a company that was not really wanted or needed but also the unhappiness of many farmers and the purchase and subsequent disposition of many dead farm animals.

This mistake occurred because of the failure to critically think through the situation thoroughly.

Had critical thinking occurred, then an attorney would have been hired to conduct a patent search and litigation could have been avoided. Also, the consultants could have provided advice concerning the prevention of marble dust entering a nearby stream. The money that could have been saved was more than offset by the desire for almost immediate payback.

The company also produced a smaller tube using the similar process. This smaller version of the tube used for the carpet was used by manufacturers of cloth, denim, material for t-shirts, and similar products. The manufacturing process was very similar. Since the tube had a smaller diameter and length, the production rate was greater. The only difference in the production process was that these tubes were manually removed from the conveyor after being heated, placed into a fixture, manually tied into a bundle of 19, 24, or 36 tubes per bundle, depending on customer specifications using a fixture and stacked onto a pallet for subsequent handloading into a trailer. This was a very labor-intensive process that was repeated once the trailer was received at the customer's dock. Various materials produced by the customer would then be wrapped around the paper tube for shipment (Figures 5.3 and 5.4).

I had just returned from a customer who produced plastic stretch wrap and had visions of grandeur. I returned to the plant and already envisioned myself in a new BMW convertible (gray, of course) that I would be driving as a result of the huge raise which I would obviously receive due to the labor savings resulting from the use of plastic stretch wrap to package tubes that could then be loaded onto a trailer while on the pallet. I spoke to the plant manager and he was in full agreement with attempting the new method. An inexpensive stretch wrap machine was purchased and delivered as well as several rolls of the most expensive stretch film available—after all

TITLE	Fixture for racking tubes for manually (orthographic view)			
Figure Number	5.3 Drawn By JD		Date	9/9/2019
Fixture for racking paper tubes for manual packaging				
Checked by JW	Date	9/20/2019 ECO #	978-005	

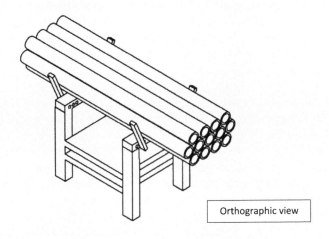

Orthographic view

FIGURE 5.3 Fixture for racking paper tubes for manual packaging—orthogonal view.

TITLE	Fixture for racking tubes for manually (plan, top, and right-hand side views)			
Figure Number	5.4 Drawn By	JD	Date	9/9/2019
Fixture for racking paper tubes for manual packaging				
Checked by JW	Date	9/20/2019 ECO #	978-005	

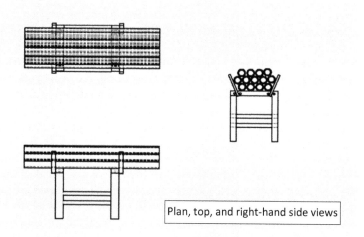

Plan, top, and right-hand side views

FIGURE 5.4   Fixture for racking paper tubes for manual packaging—top, plan, and right-hand side views.

the film was samples and free. With the help of maintenance, a fixture was designed and built (Figures 5.5 and 5.6).

All that was needed was a customer who would agree to the proposed delivery method. I met a salesman with a customer and explained the requested changes to the receiving and production departments to certify that the proposed changes would not result in any increased tasks for either department. Both departments welcomed the change since receiving palletized tubes would reduce material handling in both departments.

At this point, I looked for the number of the local BMW dealer but hesitated to call. On the next order for that customer, the fixture was placed adjacent to the oven to enable the employee to load the fixture with the tubes, placed the fixture onto the stretch machine adjacent to the fixture and the pallet with the tubes was wrapped, and the wrapped load was lifted out of the fixture. At this point, the tubes were hot having just been ejected from the oven. The plant manager, who had many years of experience in the industry, asked me if the wrapped load should be placed aside to see what happens over time. Of course, I responded negatively thinking what could happen? I responded that the load could be loaded directly into the trailer. After all, this is July in GA—what could happen—the temperature only approaches a million degrees and the humidity half a million percent? Then after the second load was wrapped, the plant manager asked again if the load should be set aside and again I

TITLE	Fixture for packaging tubes with stretch netting (orthographic view)			
Figure Number	5.5 Drawn By	JD	Date	10/1/2019
Fixture for packaging paper tubes with stretch netting				
Checked by JW	Date	10/15/2019 ECO #	978-006	

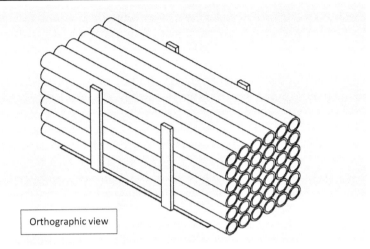

Orthographic view

FIGURE 5.5 Fixture for packaging paper tubes with stretch netting—orthogonal view.

TITLE	Fixture for packing tubes with netting (plan, top, and right-hand side views)			
Figure Number	5.6 Drawn By	JD	Date	10/1/2019
Fixture for packaging paper tubes with stretch netting				
Checked by JW	Date	10/15/2019 ECO #	978-006	

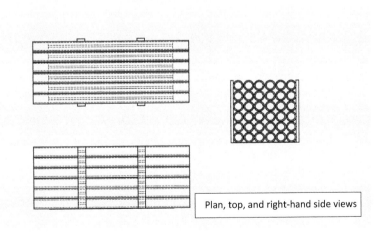

Plan, top, and right-hand side views

FIGURE 5.6 Fixture for packaging paper tubes with stretch netting—top, plan, and right-hand side views.

responded with "of course not". This load could be placed on top of the first. In fact, the entire trailer could be loaded until full. I gave him the "trust me look".

I had made arrangements to meet the salesman at the customers' dock door the next morning upon arrival of the company's truck at 7 am. When the rear doors were opened, the plans for the BMW quickly evaporated. What I saw staring at him was an absolute mess, all the once-round tubes were almost flat, caused as a result of the moisture not being able to escape due to being fully encapsulated by the plastic wrap. I thought this was not good. My BMW just drove off a cliff with myself strapped in and still screaming.

I immediately phoned the plant manager and as soon as he heard my voice he started laughing. In fact, everyone in the production office was laughing so hard they were probably crying. I told the plant manager that he had a problem and the plant manager responded with that "I messed up" in so many words. The plant manager then said, "the plant has your back—if I will turn around there is another truck with good product on it and that everyone all make mistakes—just learn from this one."

No one from the company ever mentioned that mistake, but a valuable lesson was learned and will never be forgotten. I learned to ask for advice especially from people who have more experience. I also learned that what is most important is that cost-effective changes occur if not this afternoon then as soon as possible. The company had been using that method of packing tubes for certain customers for many years, so whether a change was made this week or next was not that significant. What was significant was to make the change in the correct way.

I researched other methods of packaging and found a product called stretch netting. Stretch netting is similar to wire mesh fence but is made from plastic and stretches around the corners of a pallet without damaging the product on the pallet. Its advantage over film is that air can flow through the netting, enabling a product to be packaged out of an oven without the product being damaged.

Implementation followed a somewhat different path. I selected a different customer who would be amenable to a change, met with the salesman, incoming freight and production people, and shipped only one pallet for evaluation. Then upon positive feedback, the next truckload contained two pallets, stacked on top of each other. Upon positive feedback, then the customer received a full truckload. This piecemeal approach was employed until all customers who could be converted to this method of packaging were completed, resulting in saving freight and direct and indirect labor costs. Although stretch netting was somewhat more expensive than stretch film, significant savings still resulted.

This mistake occurred because I had not sought the advice of others. The plant manager had many years of experience and I had failed to take advantage of it. The second fact was that I did not ask any of the salesmen if they had seen such packaging from a competitor. Since it was a new method, I needed to take extra precautionary steps to guarantee success.

This mistake also occurred primarily due to the failure to employ critical thinking. The first of many lessons learned by me is that not everything works as predicted. No one else was asked because no one else had any experience with this method of packaging and my experience in it was an absolute zero. No questions were asked because critical thinking was not implemented. If it were, the results would have been as to whom to ask and what questions to ask. The second lesson to learn is that it does not have to be all or nothing—even in baseball, you get three strikes. The project could have been phased in gradually to provide learning opportunities. The third is that I had not found or made any effort to locate any company using this product to package a product similar to the one I was going to package to validate using this method. This is Engineering 101—where was I during this course or was this one of the several that I failed?

Although failure to ask others for advice is the primary reason for this mistake, employment of critical thinking would result in several additional causes for the mistakes, the misguided thinking that for some reason the change must occur now, even though the previous process has been employed for many years, and the failure to conduct proper research. This last approach is the least excusable. Over the years, I had discovered that if one talks with a company, as long as it is a non-competitor about a process that they are using, then most of them are more than willing to share information concerning it. I have had companies share costs before and after adopting the process, demonstrate the process, and be available to answer any questions that anyone representing my company may have. Most will share successes and failures as well as the main causes for successes and failures.

Failure to ask for assistance from others does not result from a lack of competence on the part of the decision maker. It usually is caused by the self-imposed pressure of the decision maker. Pressure is one of the many Human Factors that can affect the environment of work. This self- imposed pressure can be reduced by proper assignment and prioritizing of tasks and seeking assistance as needed even if higher-level team members are required [21].

These mistakes resulted from the failure to understand the importance and significance of human factors in the workplace. I encourage readers to investigate the topic further by reading the information provided under Human Factors in the Appendix section of this book and carrying out further research and coursework.

6 Conclusions and Recommendations

Resorting to the reasons and categorizing mistakes that resulted directly from the failure of the decision maker, Table 6.1 and Graph 6.1 indicate that over 80 percent of mistakes resulted from the failure of the decision maker to take some type of action.

The total number of mistakes was 33. A distribution attributed to each reason can be seen in Table 6.2 and Graph 6.1.

TABLE 6.1
Reasons for Mistakes Attributable to Failures

Reasons for Mistakes	Percentage (%)	Number of Mistakes
Failure to consider all aspects of the project	36.4	12
Failure to ask others for assistance	27.3	9
Failure to consider all costs of the operation	18.2	6
Total failures	81.8	27

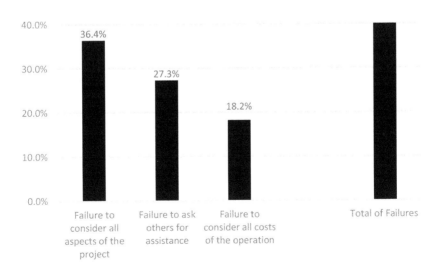

GRAPH 6.1   Reasons for mistakes attributable to failures.

TABLE 6.2
All Reasons for Mistakes

Reasons for Mistakes	Percentage (%)	Number of Mistakes
Failure to consider all aspects of the project	36.4	12
Failure to ask others for assistance	27.3	9
Failure to consider all costs of operation	18.2	6
In a hurry	15.2	5
To meet a deadline	3.0	1
Total mistakes	100.0	33

For an industrial engineer, an opportunity such as this is similar to winning the lottery even if he has to pay taxes. This is the low-hanging fruit that all industrial engineers seek.

Graphing these mistakes that resulted from the failures of the decision maker is shown in Table 6.1 and Graph 6.1 and emphasizes the significance of taking one more step before making a decision since failure to do so resulted in more than 80 percent or 4 out of 5 decisions being incorrect.

The solution is simply to get the decision maker to take one more action before he makes that final decision, to ask one more question, to get one more idea, to get one more simulation, to think through the situation just one more time to determine what is the worst that can happen if the decision is implemented and if it is the wrong one. An incorrect grade can be changed, putting diesel fuel in a gasoline tank is a costly mistake but one that can be addressed and to be sure it will occur only once.

The potential savings that could result from a correct decision versus the cost of an incorrect one will in all likelihood offset the delay in making the decision long enough to get one more input, regardless of the format of that input. In retrospect, had I simply followed my own advice and gotten an additional piece of information, many dollars, jobs, and time would have been saved. Since time travel is not possible, what is possible is to prevent or reduce the severity of errors by the application of critical thinking to obtaining at least one additional item of information before making and implementing a decision.

This is the approach that the Joint Commission employed to address wrong-site surgical procedures. The Commission issued a protocol for preventing wrong site, wrong procedure, and wrong person surgery. This protocol comprises several steps: (1) verification of the correct patient, documents, site, etc.; (2) marking of the operative site prior to the operation with the agreement of the patient so that the mark will be visible after the patient is prepared for surgery; and (3) taking a timeout to increase teamwork before surgery, which has been shown to reduce the number of incidences of wrong-site surgery [22]. The marking of the site of surgery provided additional information for the surgical team to eliminate and reduce the likelihood of an error and the increased teamwork provided by the timeout are two of the many human factors that assist in the mitigation of mistakes.

One study on refractive surgery involving almost 3,000 (2,951) consecutive patients who had a primary or enhancement Laser Vision Corrective (LVC) surgery conducted between July 2009 and February 2014, in which 1,417 (2,744 eyes) patients after the implementation of a pre-surgical checklist similar to the one from the Joint Commission, compared the number of serious errors before and after the use of a checklist. Prior to the use of the checklist, there were two serious errors at a rate of 0.07 percent. The number of errors committed after the implementation of the checklist was zero at an error rate of 0 percent, achieving the goal of a reduction in the error rate [23].

Appendix
Topics for Additional Information

ANTHROPOMETRY

Anthropometry, first developed in the 19th century, is the science of obtaining systematic measurements of the human body. It is used by physical anthropologists to study human variation and evolution in both living and extinct populations. Measurements involving the size would include height, weight, surface area, and volume; those involving the structure would include items such as height (sitting versus standing), shoulder and hip width, the lengths of the arms and legs, the neck circumference, the body composition which consists of the percentage of body fat, the water content, and the lean body mass. Various tools are used for measurements such as height, the length and circumference of body segments, bone diameter, skin thickness and subcutaneous fat, and the weight of humans [24].

Alphonse Bertillon developed a method for classifying criminal records based on anthropomorphic data, as criminal records were stored alphabetically and criminals used aliases. He began his career with the Paris Police department and is the founder of the Society of Anthropometry. He obtained measurements and later added frontal and profile photographs to each file. These photographs are still used today are referred to as "mug shots" [24].

The results are applied to sizes of clothing worn, sizes of entries and exits, capacities of elevators and escalators, and hallway ergonomics to include the tools and machines with which the employees work and the chairs and desks in which they perform the work, and medicine to include aging, obesity, and nutrition. The data is critical to prevent and reduce the severity of occupational injuries. Anthropometry studies the interaction of employees with various tasks that the employees are assigned: physical, mental, and personal protective equipment [25].

CREW RESOURCE MANAGEMENT

Crew resource management (CRM) is the process that resulted from an analysis of air crashes that indicated that more than 70 percent of air crashes resulted not from weather or equipment failure but human error. These air disasters occurred during the 1970s. The conclusion of a NASA workshop was that a majority of these failures occurred in leadership, team coordination, and decision-making.

In an effort to reduce the accidents due to human error, the aviation community responded by turning to psychologists to use tools such as group dynamics, leadership, interpersonal communications, and decision-making as a basis to develop new training tools.

The resulting training is referred to as CRM. It was explained as the use of all available resources, information, equipment, and people so that a safe as well as an efficient airline flight can occur. In addition, if properly used, it can be utilized to identify potential and existing threats and with the assistance of crew has the ability to develop, communicate, and implement plans and actions to prevent and reduce potential threats. More importantly, the employment of CRM methods has the potential of not only avoiding and managing but eliminating human errors. Employee morale and efficiency can also benefit from the implementation of CRM programs.

A safety audit was developed from the observational process at the University of Texas Human Factors Research Project to assess CRM practices. The audit, referred to as the Line Operations Safety Audit (LOSA), has been very effective since its deployment. Although no one can accurately determine the number of crashes that have been avoided or lives that have been saved as a result of the use of CRM, an analysis of LOS data concludes that errors have been observed on 82 percent of all flights with almost three (2.8) per flight. As a result of effective CRM practices, the great majority of these errors was well managed and had no or little consequence. The use of LOSA data provides organizations and regulators with a valid means of monitoring normal operations. Consequently, this information is used to determine what crews do successfully as well as where things go wrong to enable researchers to develop more effective training and safety initiatives.

As a testament to its effectiveness, CRM is being required by the International Civil Aviation Organization, a regulatory component of the United Nations, which began requiring CRM programs for member countries. CRM also informed the development of maintenance resource management, an effort to improve teamwork among aircraft maintenance workers. It is also being used by the US Air Force, among others, which now uses CRM training programs to boost communication, effectiveness, and safety among the crews that maintain and repair aircraft.

The use of CRM is being felt outside the aviation industry. The medical community wants to translate the primary benefits of CRM, improved communication and teamwork to healthcare to improve patient health and safety. CRM training is being used in air traffic control, firefighting industrial situations, nuclear power plants, and offshore oil operations to avoid making operational errors that may lead to accidents and injuries [26].

HUMAN FACTORS

The definitions will vary somewhat depending on the source, but all definitions focus on the assimilation of knowledge from various disciplines including anthropometry, physiology, physics, environmental medicine, toxicology, and biomechanics to better understand how humans perform various tasks under different circumstances. The goal of this examination is to improve efficiency, creativity, productivity and job satisfaction with the goal of minimizing errors or mitigate the severity of an accident [27].

There are several human factors that must be taken into consideration when humans are provided assignments in a work environment. These are lack of communication; the exchange or lack thereof of information, complacency; the feeling of overconfidence; lack of knowledge: lack of knowledge of specific disorders, distraction; taking the mind off what is occurring since the mind works faster than the hand, lack of teamwork; due to a large team some of whom may be new to the team, fatigue; mainly due to night shift, lack of resources, not having equipment available as needed, pressure; due to not wanting to ask questions for fear of appearing unknowledgeable, lack of awareness, not aware of what is occurring, lack of assertiveness, a degree of complacency, and norms or unwritten rules which are accepted [21].

Human factors engineering is the discipline that attempts to identify and address these issues. It is the discipline that takes into account human strengths and limitations in the design of interactive systems that involve people, tools, and technology, and work environments to ensure safety, effectiveness, and ease of use.

A human factors engineer examines a particular activity in terms of its component tasks and then assesses the physical demands, skill demands, mental workload, team dynamics, aspects of the work environment (e.g., adequate lighting, limited noise, or other distractions), and device design required to complete the task optimally. In essence, human factors engineering focuses on how systems work in actual practice, with real—and fallible—human beings at the controls, and attempts to design systems that optimize safety and minimize the risk of error in complex environments.

Human factors is therefore concerned with applying what is known about human behavior, abilities, limitations, and other characteristics to the design of systems, tasks/activity, environments, and equipment/technologies. It is also concerned with the design of training programs and instructional materials that support the performance of tasks or the use of technology/equipment. The focus of human factors is on how people interact with tasks, with equipment/technologies, and with the environment, in order to understand and evaluate these interactions. The goals of human factors are to optimize human and system efficiency and effectiveness, safety, health, comfort, and quality of life. One can apply human factors knowledge to wherever humans work [28].

INCENTIVE SYSTEMS

An incentive system is one that rewards an employee based on output. In the manufacturing environment, an industrial engineer would establish a standard time for each routing operation. Standard time has been defined as the time required by the average skilled operator working at a normal pace to perform a specified task, using a specified method, allowing time for personal needs, fatigue, and delay [29]. For an employee working alone, his pay would be computed by multiplying the number of products produced times the standard hours per product. This multiplication yields earned hours. The percentages included for fatigue and delay would vary with each individual job and must be evaluated taking all the tasks required for each job into consideration.

For an operator working on an assembly line, the controlling operation or the bottleneck on the line determines the production standard since that operation dictates the maximum production for that line. The line should be as balanced as possible to minimize unforced idle time among employees as tasks are being performed. Part of an industrial engineer's function is to equalize the work along the employees so that wait time between employees is minimized as well as work in process inventory.

Total pay is the base rate times the incentive rate percentage time the number of hours worked on incentive plus the number of hours worked at the base rate or the number of hours worked while on downtime. The number of hours worked on incentive plus the number of hours worked at the base rate must equal the total hours worked by the employee.

Incentive systems can be based on an individual or a group. Regardless of the method used, employees are paid at their base rate when they are unable to work due to a lack of work and which this period is referred to as downtime. This time must be recorded and subtracted from total hours to calculate the percentage of incentive pay earned. If a line is paid on a group incentive basis, the percentage of idle time should be constantly monitored to ensure that work continues to flow smoothly along the assembly line.

Other types of incentive pay include bonuses which are incentives that are paid to employees who exceed output goals. These incentives are in addition to their base pay rate. Commissions are another type of incentive pay. These are usually associated with sales positions as real estate or retail. Less common types are incentive pay for knowledge and skill that reward employees with higher pay as an incentive for the increased knowledge or skills acquired. Another type is profit sharing and stock option plans (PSSOP). Profit sharing is a scheme in which employers return a portion of net profit to their employees on compliance with certain service conditions and qualifications. The purpose of introducing profit-sharing schemes has been mainly to strengthen the loyalty of employees to the firm by offering them an annual bonus (over and above normal wages) provided they are on the service rolls of the firm for a definite period. The share of profit of the worker may be given in cash or in the form of shares by the company. These shares are called bonus shares [30].

INDUSTRIAL ENGINEERING

Based on my extensive experience and an extensive literature research, it is the branch of engineering that is focused on continual improvement for any processes, systems, or organization by working with people at all levels to improve and implement integrated systems of needed resources, knowledge, information, equipment, energy, and materials. This is achieved through the development of training programs, the use of time studies for direct labor and indirect labor standards and appropriate feedback for continuous improvement, adoption of effective safety programs, involvement of employees in decision-making to maximum extent possible, development of quality inspection procedures, and other programs and procedures that can affect the productivity of the company and the safety of the employees.

The Industrial Engineer today can improve processes in any industry. Many make improvements in healthcare, reducing medication errors and increasing throughput. Many work in the banking industry as can be seen in the improvements that are evident when making a deposit or withdrawal. Changes in the service logistics industry are obvious when companies offer same day service for a minimum charge or one-day service for free.

Not only do industrial engineers work to increase the efficiency and profitably of their employers, but they are concerned with the environment through the study of many related fields, such as anthropometry. For example, when designing a production line for maximum productivity, the measurements of those who will be working on the line must be considered as well as the weight of any items that must be lifted. When writing instructions, usability testing is to evaluate how well the instructions can be followed.

Industrial engineers must possess mathematical skills, since they must be able to explain the results of time studies so that they can be understood. They also need to detail a cost justification for a new or an upgrade to a piece of equipment or software. However, in my opinion, the greatest skill needed are people skills since the greatest idea in the world is useless unless it is implemented.

KAIZEN EVENT

Any action whose output is intended to be an improvement to an existing process. This is a very cost-effective method of obtaining rapid improvements to a system.

It is a management tool that

a. Assembles necessary decision makers, operators, managers, owners, etc., of a process in one area,
b. Flow charts or maps the current process and evaluates it for improvements, and
c. Encourages input and agreement with improvements from everyone involved in the process [31].

SCIENTIFIC MANAGEMENT

Scientific management could not have occurred without the invention of the steam engine, patented in 1769, which started the Industrial Revolution. Prior to the Industrial Revolution, a majority of businesses comprised a few individuals. Afterward, factories grew in the number of employees and owners and managers had little if any personal contact with employees. One of the first to realize this change that work methods need to be studied and optimized and that time studies was the vehicle by which this could be achieved was Frederick Taylor.

Taylor, a mechanical engineer, published the *Principles of Scientific Management* in 1909. He worked in situations where work was able to be quantified, systemized, and standardized. His use and analysis of time studies resulted in the redesign of shovels used in the loading of coal to feed ovens in the Bethlehem Steel Mills that enabled to reduce the number of employees performing this function from 500 to 140 or a reduction of 72 percent.

The four primary principles ar the following:

1. Examine each job scientifically and replace the "rule of thumb" method to determine the "best way" to perform the job. The "rule of thumb" enabled each employee to determine the method he desired to perform the job.

2. After the right employee was hired for a job, train them so that they could work at maximum efficiency.
3. Continually monitor their performance and initiate training as needed.
4. Separate the work so that an almost equal division of work exists between management and employees to enable so the management can manage and the employees can spend their time performing the assigned tasks efficiently [7].

Implementation of these principles occurred in many manufacturing facilities often tripling productivity. These principles were also applied in household tasks based on the results of time and motion studies. Although application of these scientific management principles benefited from the improved productivity and a significant positive effect on the industry, it came with drawbacks such as the increased monotony of work and the reduction in the variety of skills, tasks, significance, autonomy, and feedback among employees and management.

Other pioneers in scientific management were Frank and Lillian Gilbreth. Frank began his career as a masonry contractor. He developed adjustable scaffolding to keep his masons on the same level as the wall they were building, made improvements to existing cement mixers, and methods to drive pilings quicker. All of these modifications resulted from his observations of masons while working and realizing that each mason was wasting motion and his goal was to eliminate the wasted motion and energy. He studied operation and with the use of film was able to decompose each human motion into 17 elemental motions, referred to as therbligs, Gilbreth spelled backward with modifications. Therbligs was a method of decomposing the motions involved in the accomplishment of a task. The system was designed to locate and eliminate unneeded motions that resulted in the waste of time. This system was invented and improved between 1908 and 1924. Ironically, it was never the subject of any of their books [32].

After the bankruptcy of his once successful business in 1912 due to the building industry depression, he devoted more time to the study of scientific management and motion study. Through the use of a micro-motion study which he felt was an improvement over time studies, he was able to determine the motions of an employee to one-thousandth of a second. This degree of detail enabled him to document the use of his film for the operation that output could be increased more effectively as a result of more efficient use of time than by an increased speed.

Lillian Gilbreth, the mother of modern management and Frank's wife, worked with him in his consulting firm. Their stated purpose of motion study was to reduce and eliminate unnecessary fatigue which could be accomplished by the design and placement of workbenches and chairs located for the employees that could be used during regular rest periods. Dr. Gilbreth also believed that job satisfaction and indirect incentives as money were employee motivators. Together they worked to improve job and work simplification, job standardization, job simplification, a wage incentive system, and improved methods to increase employee satisfaction. Recognition of the effects of fatigue and stress on time management was a priority of Lillian and studies were made to mitigate that effect. She was the first female to become a member of the American Society of Mechanical Engineers [33].

Dr. Gilbreth, the mother of 12 children, studied Scientific Management with her husband but with a PhD in Psychology. She was more interested in the aspects of the workplace due to her keen empathy for people as well as her insight into human behavior. As a consultant, she worked for companies applying psychology to address problems in the workplace and the home. Some of her designs to simplify everyday life and make necessary work easier including the shelves inside the refrigerator and the foot pedal activated trash can. She and her husband were the parents of 12 children and she was the author of *Cheaper by the Dozen*. In addition to her consulting work, she taught college at several schools including Newark College of Engineering, Bryn Mawr, Rutgers, and the University of Wisconsin. At MIT, she was appointed a resident lecturer [34].

Another contributor best known for the development of a chart that bears his name to enable management to visually determine the progress of a project is Henry Gantt. As was Dr. Gilbreth, he was also concerned with the favorable psychological conditions that benefitted the workers. He established the quota and bonus system which monetarily rewarded those who produced more than the amount specified by management.

Still another person who made significant contributions to the field of Scientific Management is Charles Bedaux. He understood that people produced products at various rates for numerous reasons and to negate this he introduced a rating factor into time studies. Initially crude, the system has been refined over time and now plays an important role in work measurement. A rating factor for production above the normal rate would be greater than 100 percent and for that less than normal would be less than 100 percent. The greater the deviation from normal the greater the deviation from 100. Standard production is obtained by multiplying normal production by the rating factor. He also favored retaining the rest allowance as did Dr. Gilbreth to enable workers to recover from fatigue [35].

USABILITY

Usability concerns the design and the subsequent ease with which everyday products can be used and steps that can be taken to make their use easier or more user-friendly. The effectiveness, efficiency, and satisfaction are overall metrics that usability on which focuses regardless of whether the user interacts with a product, website, software, some type of device, or an appliance.

The user should be at the center of the design. Suggestions on user-centered design were enumerated and include making it easy for the user to evaluate the current state of the system, to follow natural mappings between the effect and the action, for example, turning the steering wheel to the right turns the vehicle to the right, the door to enter is on the right similar to driving on the right-hand side of the road, and to enable the user to know what actions needed to be taken at any given time. A pair of scissors is an example of a well-designed product since its use is intuitive to the user. The user can pick them up, understand the positioning of their fingers, and begin using the instrument for which it was intended intuitively [36].

Usability is multifaceted and includes among others: ease of learning which is the speed that a first time user can successfully perform basic tasks; memorability is defined as whether a returning user is able to use the site effectively; efficiency is the ease with which an experienced user can perform tasks, and error frequency and severity, which is the frequency the user commits while using the site, the severity of the errors, and the ease with which recovery can occur.

Usability engineering is a professional discipline that focuses on improving the usability of interactive systems. It draws on theories from computer science and psychology to define problems that occur during the use of such a system. Usability engineering involves the testing of designs at various stages of the development process, with users or with usability experts. The purpose of testing is to improve the product, intranet or Internet, and web site [28].

USABILITY TESTING

Usability testing is a simple process consisting of selecting a group of people, providing them an assignment, monitoring their progress or the lack of, and taking notes and time the length to completion or surrender. Organizations perform usability testing to develop and produce more usable products and it indicates that companies believe that its use is beneficial [37]. The reason for iterative design change is to modify each design as soon as possible. The testing assignment can be on a website or the assembly of an item following written instructions. If using a website, eye tracking monitoring systems can be used to follow the path of the eyes and mouse. After the task is completed or the group gives up, then the discussion needs to follow up on improvements so that the task becomes easier to accomplish. After changes are made, then another iteration of testing needs to occur with a different job assigned. This process of testing should occur until the testers agree that the product is more user-friendly.

The usability class visited a large company in the area that had a usability lab, which the class was able to observe. The company actually had some employees in the lab who were observed making changes to an employee record to ensure that changes could be made error-proof. An employee has numerous minor changes that need to be made in his record to keep the record updated as moving from apartment A to apartment B. The company believed that it was so much easier for the employee to change certain records than to take a form to HR and trust that someone in HR changes the database correctly in a reasonable time period. This was the ultimate in usability. Providing employees with this ability would increase data accuracy, morale, and provide employees some degree of control.

The class was assigned a project that consisted of working in groups of four to five students per group. The assignment consisted of designing a website, assigning a simple task to be performed on the website, and tracking the time in seconds and the number of clicks the observer made attempting to carry out the assignment. Class members were made acutely aware that they should not give up their day jobs at this point to become web designers. Then the focus of the assignment became to improve the design to reduce the time and the number of clicks by at least 25 percent. After several iterations, all groups attained their goals. Some groups required more attempts than others but all crossed the finish line.

The reason that each team was not successful was that initially a few members in any team employed critical thinking. Each attacked the issue with the attitude that websites are a dime a dozen, a piece of cake—this can be done easily—but when this task was taken as seriously as it should have been, it was realized that this was not going to be an easy, simple class task. But on analyzing the results, it became obvious that critical thinking should have occurred at the beginning of the project to ensure that the assignment is completed successfully. Although it is not a mistake, a valuable lesson was learned by all class members.

Since usability testing involves the use of humans, during the usability testing those participating must be aware of professional practices in the ethical treatment of test participants including the treatment of participants with respect, the concepts of informed consent, the knowledge that the participant can leave at any time during the experiment without repercussion, and that there is minimum risk in participating in the experiment. The guidelines also inform the participant of what will happen during the test, agreement to participate, and that participating in the test does not place the participants at any greater risk of harm or discomfort than situations normally encountered in daily life. These signed, informed forms are confidential and only the test administrator should be able to match a participant's name and data, and that data is confidential.

References

1. Huitt, W (1998). "Critical Thinking: An Overview". *Educational Psychology Interactive*. Valdosta, GA: Valdosta State University. [Revision of paper presented at the *Critical Thinking Conference* sponsored by Gordon College, Barnesville, GA, March 1993]. Retrieved 10/29/2019 from http://www.edpsycinteractive.org/topics/cognition/critthnk.html

2. Green, M. "Color Functionality in Trade Dress". *Human Factors*. Retrieved 9/25/2019 from http://www.visualexpert.com/Resources/colorfunctionality.html

3. Brodie, JM (1993). "Garrett Morgan, Inventor of One of the First Traffic Lights". *AAREG*. Retrieved 8/27/2019 from https://aa registry.org/story/garrett-morgan-inventor-of-one-of-the-first-traffic-lights/

4. American Optometric Association (2019). "Color Vision Deficiency". Retrieved 10/28/2019 from https://www.aoa.org/patients-and-public/eye-and-vision-problems/glossary-of-eye-and-vision-conditions/color-deficiency

5. Nielson, J. (2012). "Usability 101: Introduction to Usability", 3 January 2012. NN/g Nielsen Norman Group. Retrieved 6/28/2019 from https://www.nngroup.com/articles/usability-101-introduction-to-usability

6. Kohn LT. Corrigan. JM, Donaldson, MS, eds. *To Err is Human-building a Safer Health System*. Washington, DC: Committee on Quality of Healthcare in America, Institute of Medicine, National Academy Press, 1999.

7. Taylor, Francis Winslow. (1911). "The Principles of Scientific Management." Retrieved 11/2019 from https.pearsoncustom.com/wps/media/objects/2429/18.2487430/pdfs/taylor.pdf

8. Middlesworth, M (2018). "A Step-by-Step Guide to Using the NIOSH Lifting Equation for Single Tasks". Retrieved from 9/28/2019 from https://ergo-plus.com/niosh-lifting-equation-single-task/

9. National Offshore Petroleum Safety and Environmental Management Authority (n.d.). "Human Error-Failures in Planning and Execution." Australia's Offshore Energy Regulator. Retrieved 6/29/2019 from https://www.nopsema.gov.au/resources/human-factors/human-error/

10. Reason, J, Maddox, M (1998). *Human Factors Guide for Aviation Maintenance, Ch. 14* (1998). Prepared for Jean Watson. Office of Aviation Medicine, Federal Aviation Administration, Washington, DC, pp. 1–38. Revised Feb 1998. Retrieved 7/24/2019 from https://www.faa.gov/about/initiatives/maintenance_hf/library/documents/media/human_factors_maintenance/human_factors_guide_for_aviation_maintenance_-_chapter_14.human_error.pdf

11. Wojdyla, B (2011). "The Top Automotive Engineering Failures: Oldsmobile Diesels." Popular Mechanics, April 5. Retrieved 9/24/2019 from http://apps.geindustrial.com/publibrary/checkout/Alum-

12. Stefan, B (2007). "The Use of Program Evaluation and Review Techniques (Pert) in the Management of Health Organizations". *Cercetări practice şi teoretice în managementul urban. Noiembrie 2007*, Anul 2, Numărul 5S. Retrieved 9/22/2019 from http://um.ase.ro/no5S/1.pdf

13. Cooper, DW (2006). Textile and Apparel Supply Chain Management Technology Adoption. The Burlington Industries Case and Beyond. *Journal of Textile and Apparel, Technology and Management*, vol. 5, no. 2, pp. 1–22. Retrieved 10/26/2019 from https://belkcollege ofbusiness.uncc.edu/wdcooper/wp-content/uploads/sites/865/2018/05/Burlington.pdf

14. Weingart, G (2016). *Wood Explorer.* Retrieved 10/22/2019 from https://www.wood workingnetwork.com/wood/wood-explorer/sweetgum

15. Pantone Matching Systems (PMS) Definition. (2015). "What Does Pantone Matching System (PMS) Mean". Retrieved 10/30/2019 from https://www.techopedia.com/definition/487/pantone-matching-system-pm

16. Illustrated World of Proverbs (ND). "The Largest Collection of Proverbs in the World". Retrieved 11/24/2019 from www.worldofproverbs.com/2012/05/trust-but-verify-russian-proverb.htm

17. Graser, P (2016). "Copper Clad Aluminum Building Wire for Use in Residential Branch Circuit Wiring." *IAEI News*, September/October 2016. Retrieved 11/25/2019 from https://iaeimagazine.org/magazine/2016/11/10/copper-clad-aluminum-building-wire-for-use-in-residential-br

18. Bhoi, JA, Desai Darshak A, Patel, RM (2014). "The Concept & Methodology of Kaizen – A Review Paper". *2014 International Journal of Engineering Development and Research*, vol. 2, no. 1, pp. 812–820. ISSN 2321-9939. Retrieved 10/16/2019 from https://www.ijedr.org/viewfull.php?&p_id=IJEDR1401147

19. Lynn, P. (2020). "Complete Guide to Sandpaper Grit Classification". *Woodworking Nation*, Est 2017. Retrieved 3/28/2020 from https://ww.woodworknation.com/complete-guide-to-sandpaper-grit-classification/

20. NCC MERP (1996). "National Coordinating Council for Medication Error Reporting and Prevention, Recommendations to Enhance Accuracy of Prescription/Medication Order Writing". Retrieved 8/22/2019 from https://www.nccmerp.org/recommendations-enhance-accuracy-prescription-writing

21. Nzelu, O, Chandrahanan, E, Perrira, S (2018). Human Factors: The Dirty Dozen in CTG Misinterpretation. *Global Journal of Reproductive Medicine*, vol. 6, no. 2, pp. 0034–0039. Retrieved 11/15/2019 from https://juniperpublishers.com/gjorm/pdf/GJORM.MS.ID.555683.pdf

22. U.S. Department of Health and Human Services (2006). "Agency for Healthcare Research and Quality." Universal Protocol for Preventing Wrong Site, Wrong Procedure, Wrong Person Surgery. Retrieved 11/15/2019 from https://psnet.ahrq.gov/issue/universal-protocol-preventing-wrong-site-wrong-procedure-wrong-person-surgery

23. Robert, MC, Choi, CJ, Shapiro, FE, Urman, RD, Meiki, S (2018). "Avoidance of Serious, Hong Medical Errors in Refractive Surgery using a Custom Preoperative Checklist". *Journal of Cataract and Refractive Surgery*, vol. 41, no. 10, pp. 2171–2178. Retrieved 11/11/2019 from https://www.jcrsjournal.org/article/S0886-3350(15)01190-6/pdf

24. GharPedia (n.d.). "History and Basics of Anthropometry". pp. 1–8. Retrieved 4/7/2020 from https://gharpedia.com/blog/history-and-basics-of-anthropometry/

25. Hsiao, H (2018). Anthropometry The National Institute for Occupational Safety and Health, CDC. Retrieved 9/22/2019 from https://www.cdc.gov/niosh/topics/anthro pometry/default.html

26. American Psychological (n.d.). "Safer Air Travel through Crew Resource Management". *Psychology: Science in Action*. Retrieved 10/7/2019 from https://www.apa.org/action/resources/research-in-action/crew

27. Holstein, WK, Chaphanis, A "Human Factors Engineering". *Encyclopedia Britannica* Retrieved 9/7/2019 from https://www.britannica.com/topic/human-factors-engineering/Applications-of-human-factors-engineering

28. U.S. Agency for Healthcare Research and Quality (2019). "Human Factors Engineering". Retrieved 10/27/2019 from https://psnet.ahrq.gov/primer/human-factors-engineering

29. Kjell, Z (2001). *Maynard's Industrial Engineering Handbook.* 5th ed. New York: McGraw-Hill, Chapter 7.4.

30. Jahan, S (2019). Human Resource Management Practice "Incentive Systems". Retrieved 7/22/2019 from http://hrmpractice.com/incentive-systems/

31. Kaizen Event. "Definition of Kaizen Event." Retrieved 10/27/2019 from https://www. isixsigma.com/dictionary/kaizen-event/
32. Ferguson, D (2000). "Therbligs: The Keys to Simplifying Work". *The Gilbreth Network*, Retrieved 11/19 2019 from http://web.mit.edu/allanmc/www/Therblgs.pdf
33. Gilbreth, LM (n.d.). "Mother of Modern Management". Retrieved 9/19/2019 from https://www.sdsc.edu//gilbreth.html
34. Giges, N (2012). "Lillian Moller Gilbreath Biography". ASME. Org Retrieved 10/6/2019 from https://www.asme.org/topics-resources/content/lillian-moller-gilbreth
35. Kwok, ACF (2014). "The Evolution of Management Theories: A Literature Review." *Nang Yan Business Journal*, vol. 3, no. 1, pp. 28–40. Retrieved 11/19/2019 from https://www.degruyter.com/downloadpdf/j/nybj.2014.3.issue-1/nybj-2015-0003/nybj-2015-0003.pdf
36. Munari, B (n.d.). "The 10 Principles of Good Design". Retrieved 3/26/2020 from https://existentialergonomics.com/2018/06/06/the-10-principles-of-good-design/
37. Lewis, J (2006). "TR 29.3820. IBM Software Group". Boca Raton, FL, August. Retrieved 11/19/2019 from http://sistemas-humano-computacionais.wdfiles.com/local--files/capitulo%3Amodelagem-e-simulacao-de-sistemas-humano-computacio/usabilitytesting-ral.pdf

Index

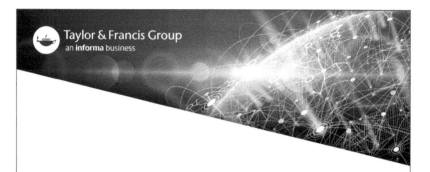